Microcomputers in Public Policy

AAAS Selected Symposia Series

Published by Westview Press, Inc.
5500 Central Avenue, Boulder, Colorado

for the

American Association for the Advancement of Science
1333 H Street, N.W., Washington, D.C.

Microcomputers in Public Policy

Applications for Developing Countries

Edited by Stephen R. Ruth and Charles K. Mann

Routledge
Taylor & Francis Group
LONDON AND NEW YORK

First published 1987 by Westview Press

Published 2018 by Routledge
52 Vanderbilt Avenue, New York, NY 10017
2 Park Square, Milton Park, Abingdon, Oxon OX14 4RN

Routledge is an imprint of the Taylor & Francis Group, an informa business

Library of Congress Catalog Card Number: 87-71155
ISBN: 0-8133-7144-9

ISBN 13: 978-0-367-00867-3 (hbk)
ISBN 13: 978-0-3671-5854-5 (pbk)

About the Book

This book presents a series of perspectives on the use of automation in the formulation and execution of public policy initiatives in developing nations. The focus is the use of the most pervasive new automation technology in the developed world—the microcomputer—as it is introduced and used in countries whose public and private sectors are just beginning to select appropriate automation methodologies. A broad range of topics is examined, including specific system development methods, appropriateness of hardware and software, painful lessons learned, successful training techniques and methods for reducing the fear associated with computer technology. There is also a section recommending Technology Sharing Organizations, quick, efficient groups whose sole purpose is to operationalize good systems techniques and implementation planning and deliver the results in months, instead of the tragically typical delay of years or more.

A surprising idea this book presents is that the best strategy for sucessful computer-based automation in developed countries is *less* money and more planning; that is, a carefully focused plan of training and detailed analysis long before programming and machine issues are considered. This "stitch in time" approach has often resulted in dramatic reductions in cost and delays in the developed world and should be used in developing nations as well.

About the Series

The *AAAS Selected Symposia Series* was begun in 1977 to provide a means for more permanently recording and more widely disseminating some of the valuable material which is discussed at the AAAS Annual National Meetings. The volumes in this *Series* are based on symposia held at the Meetings which address topics of current and continuing significance, both within and among the sciences, and in the areas in which science and technology have an impact on public policy. The *Series* format is designed to provide for rapid dissemination of information, so the papers are reproduced directly from camera-ready copy. The papers are organized and edited by the symposium arrangers who then become editors of the various volumes. Most papers published in the *Series* are original contributions which have not been previously published, although in some cases additional papers from other sources have been added by an editor to provide a more comprehensive view of a particular topic. Symposia may be reports of new research or reviews of established work, particularly work of an interdisciplinary nature, since the AAAS Annual Meetings typically embrace the full range of the sciences and their societal implications.

WILLIAM D. CAREY
Executive Officer
American Association for the Advancement of Science

Contents

Introduction

Stephen R. Ruth and Charles K. Mann

When the microcomputer first appeared ten years ago in the industrial world, large amounts of computing power already was taken for granted. Most medium and large institutions had mainframe computers; many engineers and other knowledge workers had access to them through timesharing terminals.

In the developing countries, however, there was relatively little installed mainframe computer capacity when the first microcomputers began to trickle in. For this reason, the impact and importance of microcomputers ultimately may be more profound in the developing countries than the developed. In the industrial countries, it changed the nature of access to computing power. In the developing countries, it opened access where before none had existed at all. Today's powerful and inexpensive microcomputers represent a far greater qualitative change in the public and private management environment of the developing countries than in the developed ones. In the former, there was virtually nothing; now there is an explosion of computer power; in the latter there was much; now there is more. Consequently, however great the impact in the developed countries, in the developing countries it is likely to spark even more rapid and profound changes.

This book documents and analyzes some of these changes. With trained manpower far more scarce in the developing countries than the developed, making such manpower more productive is a high priority task. In Chapter 1 Mann observes that the choice of the spreadsheet as the organizing framework or metaphor for the earliest office applications of the microcomputer made productivity gains obvious. Anyone using a pencil and paper spreadsheet could see how an "electronic spreadsheet" could speed up their work. Mann goes on to argue that while microcomputers were introduced to "automate" existing tasks, gradually the users discovered that those tasks themselves were defined in part by the primitive tools at hand. With new computer tools, new tasks—once impossible—could be undertaken. "Thought experiments" became easy with computerized spreadsheets; near real-time expenditure monitoring became practical; management information systems became timely enough to permit correcting problems before they became acute. People soon moved beyond the metaphors which facilitated their entry to computing.

Schware and Trembour appropriately inject a note of caution, urging that computers not be pushed too fast. A stark and useful contrast is evident between the Tunisian experience reported by Mann and another Tunisian computer project reported by Schware and Trembour. While both projects used Apple II computers, the former used only standard spreadsheet software allowing graduated entry to computerization. The first applications were simple extensions of familiar pencil and paper spreadsheets.

In the Central Tunisian Development Authority (CDTA) program, such off-the-shelf software was not provided. Rather, expatriates created custom programs for a new survey processing task and provided training to Tunisians on how to use it. This custom software seems to have been a critical difference. Unlike the former project where the Tunisians built their own spreadsheet models, the Tunisians in the CTDA project never seemed to achieve real ownership of the computer installation. Schware and Trembour report that "despite a great deal of technical assistance and follow-up, in the three years since the team was in Tunisia, the hardware has not been successfully installed and is not performing tasks a local or regional unit of government might find useful." This represents a sharp contrast with the many applications running on the Ministry of Agriculture's Apple II system where there was virtually no expatriate programming at all, but where two Tunisians were given a six week course in using standard software packages. This contrast suggests the importance of avoiding the "not invented here" syndrome when introducing computers. It suggests the usefulness of "user friendly" programs where local staff can set up their own procedures based upon training in general purpose software rather than custom programs created from scratch.

Schware and Trembour go on to present an insightful appraisal of the Egyptian National Health Information System. One of their observations is that part of the computer's benefit was the discipline it imposed on office systems to "get ready for the computer." The benefits of this computer-inspired discipline also were reported by M. Boujbel in the Tunisian MinAg project. Schware and Trembour go on to draw some useful recommendations concerning microcomputer installations. These lead into more general treatment of that topic by Gotsch, a pioneer in the use of microcomputers in agricultural development applications. (It was at his training program at Stanford where the two Tunisian technicians learned their skills.) Drawing upon his experience with microcomputer installations, he sets forth some rules of thumb to help identify circumstances in which the application of computers is likely to be successful. He also discusses particular environments and tasks to which microcomputers are particularly well-suited. He concludes with observations about system hardware, software, training, maintenance, and institutionalizing microcomputer capacity.

Palmedo's chapter deals with energy planning, illustrating with examples that the microcomputer changes not only how analysis is carried out, but what analyses are performed, the same point observed in the MinAg project in Tunisia. Similarly, he notes the importance of the sense of the application being "invented here." "As national staff have more direct input and participation in the analytic efforts the results are achieving greater support."

In contrast to other chapters which include sections on training to use microcomputers, Palmedo calls attention to the usefulness of computers in training people; computers, for example, running simulations of heat balances in a building as control variables are manipulated. He notes that "microcomputers are now used extensively to teach energy planning and analysis techniques in the Energy Management Training Program at the State University of New York at Stony Brook." As to the training for computer use, he challenges the conventional practice of selecting hardware, then software, then training the staff to use it. Exactly the reverse emphasis is required, he believes, with priority attention going to the training component and other considerations flowing from this emphasis on the human element.

The Pinckney-Cohen-Leonard and Wescott chapters document and analyze the enormous impact which the microcomputer has had in several government ministries in Kenya. Only four years after introduction for budgeting in the Ministry of Agriculture, microcomputers were used for the production and publication of the entire national budget. Pinckney, Cohen and Leonard describe some of the problems which faced Kenyan officials, and how the microcomputer helped to overcome these problems. Not only did anticipated benefits materialize, but some unexpected benefits occurred as well. The long association of the authors with this particular project allowed them to formulate a useful set of generalizations: lessons learned over the course of the project. These include considerations concerning institutionalizing use of the equipment. Some of the greatest hurdles involve personnel issues, tending to validate the priority on human factors suggested by Palmedo.

In effect, Wescott picks up the story where the Pinckney–Cohen–Leonard piece ends, as the Ministry of Finance adopts microcomputers for budgeting following the early successes in the Ministry of Agriculture. His chapter is particularly effective in stressing the interactions between the technology and policy, most notably the role of the computer in facilitating implementation of the government's decentralization effort. Wescott includes a blow–by–blow description of the introduction of computers into the budgeting process, included to demonstrate some of the pitfalls which await the unwary. Like the others, he highlights the main lessons learned in an excellent section, "Principles in Introducing Microcomputers."

In general, these principles are consistent with the observations of the preceding authors, although he goes somewhat beyond them in several respects. He urges, for example, that when computerizing budgets, "begin, if possible, with the Ministry of Finance." He raises to the level of a general principle the observation of others that "Many of the benefits of microcomputers will be unexpected." He calls particular attention to the important role of senior management and simultaneously to a core group of enthusiastic promoters at the grassroots level. He calls attention to the "new enthusiasm...from mastery of powerful new skills. This enthusiasm is quickly transferred to senior management, who discovers a new, hardworking resource that delivers what it promises on time." There are echoes of the Tunisian lessons in his admonition that "the operation must not become a foreign one."

Brodman's contribution draws both upon the Kenyan projects reported above and an Indonesian project to highlight the relationship between management factors and the impact of computers on decision–making. Perhaps her most significant contribution beyond the previous authors is her explicit focus on incentive systems and "the power structure." She concludes that "management systems and structures, rather than technical issues of hardware, software, and computer skills, emerged as key to determining the response to microcomputer adoption...." Like Palmedo, she calls attention to "the importance of providing training in the *use of information*, not just training on how to produce and manipulate it" (emphasis hers). Her vantage point as a political economist helps to set all of the case examples in broader perspective.

One way or another, all of the authors note the importance of the developing countries making computer projects their own rather than viewing them as a foreign transplant. Render and Ruth carry this reasoning still further, urging that developing countries examine the software industry for areas where they may have a comparative advantage in producing software. All the other contributors write in terms of developing countries solely as consumers of high–tech, not as participants in its production. It is, they state, "not too late to join this new international industry." They propose concrete steps by which such development can be encouraged.

Next is Ruth's proposal for making more effective use of technology sharing organizations (TSOs). These would be structured and staffed to address problems in the developing countries which fall outside the range and/or scale of activities of present organizations. Probably they would be affiliated with existing institutions or professional associations, but with this technology transfer mission made explicit. The chapter advances "a modest proposal to unite the capabilities of existing scientific organizations with the need for some regions to solve smaller scale development problems." The TSO could become one vehicle by which the observations, principles, and experience decribed in the preceding chapters could be disseminated more widely within the developing countries.

The volume concludes with some final observations by Mann. He notes that much of the initial application of the microcomputer was in automating a process already familiar and being done by other means. The real promise and, in a sense, the risk of the microcomputer may lie not in what it is doing now, but what it may do in the future, particularly as more "intelligence" becomes incorporated in software. Expert systems in particular appear to have a bright future in helping to raise the productivity of the relatively few well–trained developing country scientists, engineers and health professionals. Much of the promise of the microcomuter lies beyond the metaphors which underlie the first generation of microcomputer programs.

1

Beyond the Metaphor: Microcomputers in Public Policy and Human Capital Development

Charles K. Mann

Introduction

A central objective of many technical assistance projects in the developing countries is increasing human capacity, defined in Webster's New Collegiate Dictionary as "the power to grasp and analyze ideas and cope with problems." Despite this focus on capacity, there has been little effort to explore systematically by what means capacity can be improved. In particular, outside of those looking at computers in schools, almost no one has explored how those most powerful tools of mental productivity—microcomputers—relate to the development of human capacity.

Often in a very intuitive way, microcomputers have been introduced into many development projects with great success—witness the case studies in this volume. Their impact on institutional performance, on various aspects of national planning, on individual career prospects, on employment, all are addressed in this book and others. Yet few authors address in any really fundamental way how the use of the microcomputer affects the development of human capacity—the ultimate

---

An earlier version first appeared as Development Discussion Paper No. 235 of the Harvard Institute for International Development, Cambridge, Massachusetts, November 1986.

objective of so many projects. Ironically, it is a computer company's advertising slogan that seems to come closest to the mark: "We change the way the world thinks." While the company no longer cares to be identified with this slogan, there is a grain of truth in this statement about computer use. It behooves those of us working in projects to develop human capacity to come to grips with this phenomenon in both its positive and negative aspects.

This chapter represents an attempt to stimulate discussion on this capacity development dimension of microcomputers in development. It draws upon my experience with a Rockefeller Foundation project in Tunisia and the experience of my Harvard colleagues in Kenya as reported in Chapters 5, 6, and 7.

Context of the Projects

Turning first to Tunisia, in 1980, the government there asked The Rockefeller Foundation for assistance in assessing the causes of a drastic decline in cereal production over the preceding three years. An incidental part of that help was two Apple II computers with VisiCalc "electronic spreadsheet" software. Rapidly, however, these two machines became the focal point of the project. They have transformed how the agricultural planners think about the problems, where they seek solutions, how they go about their work.

In Kenya, falling coffee prices and contracting world markets precipitated a serious financial crisis in the early 1980s. The budgeting and expenditure control process which had worked in an era of rising revenues could not generate the timely reports needed to manage resources more frugally in a time of austerity. Moreover, the International Monetary Fund (IMF) had applied pressure for more effective and detailed monitoring of financial and monetary aggregates. Additionally, a decision to decentralize financial authority required an expanded ability to disaggregate financial and project information to the district level. These pressures led to the introduction of microcomputers into Kenya's planning and budgeting system.[1] The resulting improved financial management has given them an increasingly important role, culminating in 1985 when the full government budget was produced using a microcomputer system.

Both The Rockefeller Tunisia project and the HIID's work in Kenya are aimed fundamentally at helping the two governments to strengthen managerial and analytical capacity within key government offices and ministries. It is becoming increasingly apparent that microcomputers are making an important contribution to this process; that there is an important interaction between the human and physical capital involved—people and the microcomputers.

In reflecting upon the lessons of these two projects, this chapter seeks insight from some of the basic principles of cognitive psychology, applying them to understand better the observed responses in Tunisia and Kenya as microcomputers were introduced. These principles help shed light on the role of this technology in the development of managerial and analytical capacity.

To establish a context for the discussion, let us begin with a vignette from the Tunisia project. The initial impetus for introducing computers in the project was the simple observation that much of the planners' time was consumed in periodic pencil and paper recalculation of large numerical spreadsheets. The computers could facilitate doing existing tasks better and faster. This was deemed sufficient justification for adding them to the project. Little thought was given to new things which might be done with the equipment. However, as the project evolved, the new uses and their associated side-effects came to be more important than the tasks which originally motivated adding the computers to the project.

The minister of Agriculture one day asked his chief of Agricultural Planning, M. Mohsen Boujbel, whether with his "miniordinateur" he could project when Tunisia would be able to reach self-sufficiency in cereals. With the base information already in VisiCalc spreadsheets, M. Boujbel recounts that:

> We went to our microcomputers and we gave... the response that afternoon because we had the demand projections and we had different alternatives of yield and of area. We produced the balance sheets very quickly and we wrote him a paper in two days.[2]

The forecast was grim. Earlier views held that with good rains, the problem would ease. The projection made it obvious that even with the best growing weather, current levels of effort would not close the import gap. Quantum change was needed. Staff members credit this projection exercise with creating at senior management levels a new sense of urgency to improve agricultural productivity, to increase the use of fertilizer, weed control and new seed, to accord generally higher priority to agricultural investment.

With no outside help, the Tunisians had created this view of the future, this model whose structure and assumptions they controlled and manipulated. Its impact in reordering priorities was far more profound than that of the more complex models earlier created by a team of foreign experts. The Tunisians, using a system they understood, had constructed their own meaning from the data and by so doing were prepared to act upon the results. The dire predictions of the foreigners' model could be more easily dismissed as the abstract and theoretical construct of people who did not really understand Tunisia. This one, however, was their prediction and it galvanized action. The computer facilitated constructing a framework within which the data could be systematically manipulated and examined.

Insights from Cognitive Psychology

The Role of Framework

Seeing knowledge in the context of an organizing framework helps both remembering and understanding. Psychologist Jerome Bruner, for example, notes that "the most basic thing that can be said about human memory, after a century of intensive research, is that unless detail is placed into a structured pattern, it is rapidly forgotten."[3] Further, he states: "The teaching and learning of structure, rather than simply the mastery of facts and techniques, is at the center of the classic problem of transfer."[4] On one level, the framework may be the fundamental principles which structure the knowledge of a particular discipline. An example from agronomy would be Liebig's Principle of Limiting Factors. This says that plant growth will be restricted by whatever nutrient the plant runs out of first, no matter how much of the others are present. This principle provides the framework within which agronomists organize much of their work.

On a more superficial level, a framework may be a more mechanical device such as the rows and columns of a spreadsheet which facilitate categorizing and organizing information. It may be a mathematical model embedding certain relationships of some quantities to others. In the case of an "electronic spreadsheet" program such as VisiCalc, the mechanical and the mathematical can be combined. The power of the computer allows the user to go far beyond what can be done with pencil and paper. Once the rules of using the framework are mastered, new uses begin to suggest themselves. "Why, if I can do this, then also I should be able to do that, and if I could do that, then I could do even...."

An excellent example of such a combined conceptual and mechanical framework is given by Gotsch in his discussion of the role of "macros" (Chapter 3). These are procedures enabling a computer user to call up with a single keystroke a whole sequence of operations within a spreadsheet matrix. A standard conceptual framework for thinking about investment decisions is benefit–cost analysis: a systematic method of comparing benefits with costs adjusted to account for the time value of money and other factors. For all the simplicity of the basic idea, as Gotsch notes, "the implementation of benefit–cost analysis for investment projects is full of conceptual pitfalls. Many of these can be avoided if the calculations that are required are already embedded in a spreadsheet template."[5] On the one hand, this means that relatively less trained personnel can perform the analysis using these prepared spreadsheets. On the other hand, as the designer of the analysis embeds successive levels of complexity into the framework, he or she is freed to think about further refinements at yet higher levels of sophistication. Using a military metaphor, it allows one to consolidate the ground taken and to contemplate the next advance. Without such embedded routines, the same ground must be retaken with every new project. This progressive mastery of complexity is what encouraged the Tunisian planners to move from automating their current statistical tables to forecasting their cereal demand/supply balance in the year 2000.

The Role of Metaphor

A powerful way of learning new information is to relate it metaphorically to information already known. Doing A is like doing B; X is like Y. Increasingly, the use of metaphor in learning is recognized as a way of tapping the power of the "right brain" to assist learning. Williams, for example, speaks of two modes of thinking; one linear, one holistic. The first is associated with the left brain, the second with the right.

> In place of a list of separate attributes, it [metaphoric thinking] offers a single image which contains most of the attributes. Like the containerized boxcars that can move from ship to train or truck, it enables you to deal with a single whole instead of its many constituent parts. The metaphor also creates a link between the list of attributes and your own experience.[6]

Metaphors have been important in helping people understand how to use microcomputers. The spreadsheet metaphor has been a powerful vehicle for understanding an important class of uses for the microcomputer. It was seeing large sheets laboriously being filled in and periodically changed with pencils and erasers which prompted the original suggestion to introduce the Apple II system into the planning office in Tunisia. It was conceived as a tool to automate procedures already in use. This it did, but had it done only this, it would have had an impact only at the clerical level. No one had thought about using a pencil and paper spreadsheet for a model within which to manipulate numerous variables, growth rates, and ecological parameters.

The clerical staff had trouble enough just keeping up with current information.

The beauty of the spreadsheet metaphor was that it provided an immediate and powerful insight into the new computer technology. The drawback was that initially it constrained thought to the limits of paper and pencil spreadsheets. If it had been only the clerks learning how to automate what they were already doing, management might never have seen the new world of possibilities which opened with the ability to enter complex formulas into the spreadsheet cells. However, management was actively involved. They came to see possibilities far beyond the metaphor. Altogether different things could be done with the electronic spreadsheet than could be done with a paper one. As Lemmons puts it:

> As you watch the effects of changing one figure ripple through a whole model, you are able to think in larger terms than a calculator and paper allow. Furthermore, the visual rippling itself sometimes is tantamount to a graphical simulation of the problem under study. You gain a better understanding of relationships in the model and a feeling for the probable consequences of different kinds of changes. When a change causes startling results, you realize that either the model doesn't reflect reality or the reality is different in some important way from what you thought. The spreadsheet provides a ready means for analyzing the startling result.[7]

It was only as the Chief of the cereals planning group began to work with VisiCalc that he began to realize the potentials which lay beyond the metaphor.

In the Kenya case, as well, it appears that metaphor was important in helping people relate to the computerization of operations. Unlike Tunisia, where the spreadsheet was the principle focus of the computer application, in Kenya this format proved inadequate as the basis for an expenditure monitoring system. While it represented a great improvement over the previous manual system, it became clear that a database management program would bring still further advantages. Here too, introduction seems to have been facilitated by metaphor. In describing the introduction of a dBase II (database management) system, Pinckney, Cohen and Leonard note that the system design "allows the programmer to set up data entry screens that look like the printed form."[8] In introducing clerical workers to the new system, this superficial resemblance permitted linking the computer metaphorically with something already familiar. Again, the metaphor provided the point of entry, but the system provided the power to move far beyond this starting point. For example, while the screen may look to the clerk like the form it replaced, now "a number of validation routines check the data as it is entered to improve accuracy."[9] Once more the user moves beyond the entry-level metaphor.

In Kenya, as in Tunisia, system start-up was facilitated by linking it metaphorically with the manual system it was to replace. The same point could be made about the equipment itself and "the typewriter metaphor." Indeed, one of Apple's most brilliant marketing strokes was making the Apple II look like a typewriter with a television and not like a computer console. It provided a strong identification with the familiar rather than the suggestion of something fundamentally different. On the one hand, the typewriter metaphor facilitates the approach to the computer. In particular, it provides easy entry to word processing. On the other hand, it tends to blind the uninitiated to the enormous range of capabilities of word processing for which there is no typewriter analog. Moreover, in the developing countries, it tends to focus attention on the cost of potential displacement of typists rather than on the benefit of making more efficient use of scarce managerial and analytical talent. Some sense of this benefit is apparent from Brodman's description of the pre-microcomputer budget process in Kenya:

> Once the figures were calculated, they had to be typed by secretaries. The document then had to be proofread by the finance officers. The number of typing errors usually necessitated the retyping of several pages, which then had to be completely re-proofread. Finance officers and secretaries often went through the typing-proofreading cycle several times. The document then was typeset, requiring yet another cycle of proofreading, corrections, and further proofreading.[10]

Where skills are a constraint to development, this conservation of scarce finance officer time may be more significant than any displacement of keyboarding labor.

Individual Responsibility for Learning

A central principle of cognitive psychology is that learners do better when they "believe that they [are] origins—people who cause their behavior—rather than pawns—people largely under control of their environment or of other people."[11] In effect, people who take some responsibility for their own learning are able to learn more effectively. The microcomputers allowed the Tunisian planners to take greater control in attempting to understand their agricultural situation. Contrast the situation of the planner designing a VisiCalc model to forecast production in 2000 to his being given a complex printout of an expatriate's complicated mainframe computer model. The whole learning situation was under local control and the results more powerfully internalized.

The same power of seeing oneself as the origin of change rather than a pawn is evident in the Kenya situation. Some key Kenyans participated actively in the design of the system and understood the basic structure of the new database. It was not just some unknown black box installed by the foreigners. The importance of this is suggested in the following observation by Pinckney, Cohen and Leonard:

> The presence of the microcomputer visibly changed the technology of financial decision–making. Officers knew that it was possible to do things that had not been done before. While some of the reforms... might have been possible without the machines, it would have been harder to convince officials that they were desirable, possible, and worth the effort. Similarly, the microcomputers legitimized a role in financial decision–making for those who could use it, therefore making it easier for certain reform–minded individuals to participate actively.[12]

The Principle of Constructed Meaning

This represents another important principle of cognitive psychology. It offers another key to understanding the powerful interaction between the computer and the person who decides what he or she would like to have it do. Quoting again from research results summarized by Wittrock:

> ...We mentally construct the reality in which we live. The human brain constructs a representation or model of reality (Jerison, 1977). Rather than reacting automatically to our environments, we use the information it supplies along with our schemata and our history of previous learning to construct meaning out of nearly any ambiguous or well–structured situation.... The processing people perform upon information they receive, as well as the information itself, influences learning, probably by allowing the learners to construct relations among their distinctive memories, previous experiences, schemata, and the incoming information....[13]

Continuing, Wittrock draws out implications for learning from this principle that we generate our own reality.

> One of these implications is that learners should be induced to construct mental elaborations of the materials they are

> trying to learn. Analogies, similies, and metaphors may sometimes help them to construct verbal elaborations; and pictures, graphic models, concrete examples, and concrete words may sometimes help them to construct imaginal elaborations.[14]

Wittrock goes on to note another implication of the principle: learners differ in the mental processes best for each of them to use to learn different kinds of information. Nevertheless:

> In either case, the facilitation of understanding and memory involves active cognitive processing of information, especially the active generation of relationships between different segments of new information and between past learning and new information.[15]

These insights from cognitive psychology help to explain why the "reality" generated with VisiCalc could influence action so powerfully in the context of the Tunisian project. First, the results were not handed them by expatriates from some model over which they had little control nor much understanding. Second, the VisiCalc model represented a reality which they generated; into which they fed their knowledge so that the future state related to what they knew in a way that they specified.

The computer not only produced the above modeling results and many other kinds of useful output, but it proved to be a powerful tool for capacity development, the central objective of the project. It provided a mechanism for casting problems in a usefully structured way; it facilitated the integration of new information with old; it suggested new and more productive ways of thinking about problems; it expanded the perception of what was possible; it extended the range of factors over which the planners believed they could exert control. Moreover, as will be indicated, it had a number of organizational benefits. Not only did it improve capacity at the level of the individual, but of the organization as well.

The Principles in Relation to the Tunisian Project

Practicing what it preaches, this chapter itself uses the framework provided by some fundamental learning principles to construct meaning about a project involving microcomputers to make sense of observed outcomes. (Both because the Kenya experience is relatively well-documented[16] and because of my first-hand experience with the Tunisian project, the balance of this chapter focuses on the Tunisia case.) In light of the principles set forth above, it examines the impact of the microcomputers upon what was done within the project, how it was done, and what secondary effects occurred. The main information source is an in–depth interview with Mohsen Boujbel, Tunisia's Chief of Agricultural Planning.

As suggested by the example which begins this chapter, one of the main uses of the microcomputer in Tunisia has been to create simple models which can help to simulate the effect of various policy changes. M. Boujbel mentioned specifically forecasting production under various assumptions of inputs and weather, estimating the effects of altering credit policies and changing prices and subsidies. An anecdote on this latter point is particularly revealing. In 1983, bread subsidies in Tunisia were eliminated, causing a sharp rise in the price to consumers. Serious food riots resulted with substantial loss of life. I asked M. Boujbel if the computers played a role in the analysis leading to this decision. He responded:

> When the government doubled the prices, there were a lot of problems here in our country. The President had said that he wanted to have the prices less than those announced, and he asked the government to do the calculations and the incidence. They answered that they cannot do that in one night because it is a lot of data processing; but, I said to my boss, if the Ministry said that to me, I said that if they had these microcomputers and the models, it then is a question that's not complicated. I said that if they had these kinds of computers, the answer could be done in two hours, not more.[17]

With the ability to calculate rapidly the effect of alternative changes, the decision makers would not have faced such an all–or–nothing alternative as was presented. Perhaps, suggested M. Boujbel, the riots could have been forestalled.

In another observation on the effect of microcomputers, M. Boujbel cited the discipline which the VisiCalc framework provided for the collection of data, and the way this speeded up their ability to advise their Minister on policy questions.

> The problem is that the decision makers are always wanting to have very quickly an answer; they haven't time to wait. They must have the decision here, a decision there... and the economists, the engineers, they say, 'Oh, we cannot have the answer, we must wait for one month or two months for work of this kind,' and the decision makers haven't time— sometimes the problems cannot wait. [The microcomputer] allows them to get back very quickly with the answer and to have, in my opinion, the less bad solution. It is very important sometimes to have the less bad solution because you are

> not in a country with many, many data in your hands, and you must have really a rough opinion, sometimes, on very few data.... The microcomputers can help you, can push you to have the data you need. You have the framework, the tables, and you want to fill in the figures and it takes some discipline to put in these figures. It's what I want to do with my people, this department, to have very disciplined gathering of data on each subject... on credit, on extension, on irrigated area, on many subjects, because you are working on many areas and we are supposed to have the answers on different questions.[18]

We see in M. Boujbel's statement the manifestation of the power of the framework to focus and aid thought and the way that the computer helps to suggest, to generate, and to implement such useful frameworks. Further, in a world where the academic literature constantly presses for the optimum, it is refreshing to hear a senior official candidly stating that he would be happy simply to have a less bad solution.

The indirect impact of the computers on the work of the Planning Office can be grouped into two broad categories: effects on efficiency and on the internal and external relationships of the office. On efficiency, M. Boujbel observed that the microcomputers give "people and engineers more time to analyze the figures," rather than taking most of their time for the calculations themselves. Moreover, "it lets us do more things than before... and it takes us really very little time to do more because it responds very quickly."[19] Nor is the microcomputer use limited to specialists: "It can be accessible to people who have never heard about computers before."[20] In addition, in an environment where computer power has been available only in large centralized systems, the decentralizing potential of the microcomputer opens new possibilities. Again, in M. Boujbel's words:

> It's very important, this side of the microcomputers; it lets you go in on the microcomputer any time you want, you are not in a very strict system where you have to come at this time, because there is not enough memory. You have your own machine, and I think in the developing countries, if we arrive to use all the capacities of the microcomputers, we will perform a lot better in our decisions, really![21]

Internal working relationships improved as the computer provided a focal point for cooperation. Drawing upon the power of metaphor, one can liken a microcomputer in an office to a campfire at a picnic. Not only does it do useful work, but it catalyzes social interchange, helps create an atmosphere of sharing information, provides a centering point for the activities of the group. The negative potential also is captured by the metaphor. If people lose interest in the campfire, no longer will it warm them. Conversely, if it is improperly tended and maintained, it can also destroy things of value to the group. The positive side of this "Campfire Effect" is evident in M. Boujbel's observations: "It brings all the engineers and the economists together as a group and they want to learn... how to use, to perform themselves with the microcomputer. They develop a very good relationship among themselves because they ask many questions, one to the other. It helps the maturity of the group."[22]

In addition to improving relations within the office, it strengthened the office's relationship with the Minister and with other departments as well. Speaking of his "Briefing Book," M. Boujbel noted:

> When we are now with the Minister of Agriculture, sometimes we show him all the figures and how the [agricultural] sector is going; after one year, we can say that the response is there. He asked us some questions, we answer, and he's enthusiastic about the microcomputer because he wants quickly... to have the implications for all the policies we study. The microcomputer allows this.[23]
>
> We can have the attitude in our administration for the decision makers to ask the economists, and they also can use the machinery very quickly and have the dialog flow in the system.[24]

Horizontal as well as vertical relationships improved. Again, one can see in M. Boujbel's comments evidence of the Campfire Effect, this time drawing those outside in towards the warmth and camaraderie of the fire.

> The computers serve us as a bridge with other departments because when we went to a meeting, we presented our inputs on the microcomputers. It leaves our department open to other departments and to the University, the College of Agricultural in Tunis, the Institute of Quantitative Economics of the Ministry of Planning. It is a really good bridge for us.[25]

Unanticipated consequences were not limited to the agricultural sector. The microcomputer is endowed with the mystique of science and its promise to improve the lot of the developing countries. M. Boujbel related:

> This friend of ours asked us to take all the equipment to his native village for the teachers of the young people in the primary school and secondary school to show them how the

> United States and the developing country people are working and how the computers can really help and to open them up to this kind of science. He wants really to have a microcomputer club in the little village.... I think that this will be a very, very good thing because if we can start some clubs in our country, we will have many young people working, training themselves on these kinds of machines. And it will come as a demystification of the computer science, you see. It is very important....
>
> Mr. [President] Bourgiba's son was very impressed when he went to the United States when he realized how widespread is the usage in the United States of the microcomputers.... [Upon his return, he] went to Algeria and to Morocco and he discussed with the President of Algeria and the King of Morocco having a center for microcomputers and of computers in Tunisia, and the Ministry of Agriculture gave them land to install it in a very nice place, this center of microcomputers. They are really pushing and believing in this kind of machinery and science.[26]

This vision of a regional center for "Informatics and Telecommunication" is now taking shape with some assistance from US AID.

Conclusion: Beyond the Metaphor

Because the microcomputer is accessible through metaphorical association with the familiar—the spreadsheet metaphor, existing data forms, the typewriter—it can be assimilated quickly and easily in a developing country environment. However, were it only to automate these familiar applications, it would not come to dominate project evolution as it did in the cases described earlier. In Tunisia, for example, calculators were also part of the project. While they significantly improved efficiency, they did not cause changes in the way people conceptualized the problems they faced. The microcomputers, in contrast, did catalyze fundamental shifts in perceptions and in organization. While the computer was accessible at entry level through metaphorical association with the familiar, people discovered that they could move far beyond the metaphor. It serves only as a starting point. It is this power to go beyond what they are already doing which has captured the imaginations of managers in the developing countries. This is the promise which the microcomputers hold for them. It is the combination of accessibility through metaphor with the ability to carry the user far beyond it that makes the microcomputer an appropriate technology in many developing country situations.

In designing projects with microcomputers, we too must look beyond the metaphor. The typical "needs assessment" focuses on what already is being done, in effect using current practice as the metaphor. Adding a computer is conceived simply as automating the process or activity rather than the means of transcending the limitations of current practice.

New generations of software which embody expert knowledge will permit fundamentally different approaches to such age-old challenges as crop management. For example, an expert system linked to a simulation model of the cotton plant is now under test at the farm level, running on an IBM PC or equivalent. Based upon actual and predicted weather, crop and field conditions, the Comax system (COtton MAnagement eXpert) determines optimum fertilizer and irrigation application and predicts the ideal date for applying pre-harvest chemical treatment.[27] Where there are few experts, embodying such technical knowledge within inexpensive microcomputers may represent one avenue to the next Green Revolution. Yet such applications lie far beyond the familiar metaphors which have carried the first microcomputers into the developing countries. It is in this world beyond the metaphor where the true promise of the microcomputer lies. Yet it is access through metaphor which helps make it achievable. The metaphor must serve only as starting point, not as end point.

Notes

1. Janice Z. Brodman, "Microcomputer Adoption in Developing Countries: Old Management Styles and New Information Systems," (College Park, Md.: Prepared for the International Development Management Center, University of Maryland with funding from the U.S. Agency for International Development, 1985), mimeo, p. 16.

2. Mohsen Boujbel, "A Planner's View of the Microcomputer," Transcript of an interview with Charles K. Mann, "The Micro Circle", mimeo, 1984, p. 5.

3. Jerome S. Bruner, The Process of Education (Cambridge: Harvard University Press, 1960) p. 23.

4. *Ibid*, p. 12.

5. Carl H. Gotsch, "Applications of Microcomputers in Third World Organizations," Chapter 3, this volume, pp. 41, 42.

6. Linda Verlee Williams, *Teaching for the Two-sided Mind; A Guide to Right Brain/Left Brain Education*, (Englewood Cliffs, N.J.: Prentice-Hall, 1983), p. 57.

7. Phil Lemmons, "Beyond the Word Processor," *BYTE*, January 1984, p. 53.

8. Thomas C. Pinckney, John M. Cohen, and David K. Leonard, "Kenya's Introduction of Microcomputers to Improve Budgeting and Financial Management in the Ministry of Agriculture," Chapter 5, this volume, p. 75.

9. *Ibid.*

10. Brodman, *op. cit.*, p. 62.

11. M. C. Wittrock, "Applications of Cognitive Psychology to Education and Training," in *Cognitive and Affective Learning Strategies*, Harold F. O'Neil, Jr. and Charles D. Spielberger, (eds.) (New York: Academic Press, 1979), p. 310.

12. Chapter 5, p. 80.

13. Wittrock. His reference to H. J. Jerison is: "Evolution of the Brain" in M. C. Wittrock *et al.* (eds.), *The Human Brain*, (Englewood Cliffs, N.J.: Prentice-Hall, 1977).

14. *Ibid.*, p. 312.

15. *Ibid.*

16. In addition to the chapters in this volume: C. H. Gotsch, "An Evaluation of the Present Contribution and Future Potential of Micro Computers in the Kenyan Ministries of Agriculture and Livestock Development," mimeo, 1983; D. K. Leonard, J. M. Cohen, and T. C. Pinckney, "Budgeting and Financial Management in Kenya's Agricultural Ministries," *Agricultural Administration*, XIV, (2), pp. 105–120; T. C. Pinckney, J. M. Cohen, D. K. Leonard, "Microcomputers and Financial Management in Development Ministries: Experience from Kenya," Development Discussion Paper No. 137, (Cambridge, Massachusetts: Harvard Institute for International Development, 1982); Clay Wescott, "Microprocessors and Financial Management for District Focus in Kenya," mimeo, 1983. On Tunisia, for a comparison of two Apple II installations, see Charles K. Mann, "Improving the Productivity of Food Policy Analysts: A Tale of Two Computers," in Charles K. Mann and Barbara Huddleston, (Eds.) *Food Policy: Frameworks for Analysis and Action*, (Bloomington: Indiana University Press, 1986). For a popular account of the project: Andrew Christie, "Apples in Tunisia: The Problem of Wheat," *Softalk*, May 1984, pp. 120–122.

17. Boujbel, *op. cit.*, p. 6.

18. *Ibid.*, p. 7.

19. *Ibid.*, p. 2.

20. *Ibid.*, p. 7.

21. *Ibid.*, p. 6.

22. *Ibid.*, p. 4. Rockefeller Field Staff member Ms. Christopher Mock assisted the Tunisians in setting up an informal microcomputer training program. She noted in this context how the shared experience of learning computer skills helped to build good working relationships across organizational boundaries.

23. *Ibid.*, p. 3.

24. *Ibid.*, p. 6.

25. *Ibid.*, p. 4.

26. *Ibid.*, p. 6.

27. Hal Lemmon, Comax: An Expert System For Cotton Crop Management, *Science*, Vol. 233, 4 July, 1986, pp. 29–33.

2

Rethinking Microcomputer Technology Transfer to Third World Countries

Robert Schware and Alice Trembour

This chapter discusses the problems of introducing microcomputers into Third World countries, with emphasis on the importance of ensuring the accuracy of the manual data collection system.

In the course of the recent proliferation and rapid diffusion of information technologies in First World countries, computer technology has begun to spread to most Third World countries (TWCs). One result is the coining of the term "barefoot microchip," a throwback to earlier concepts of development strategies, such as the "barefoot" doctors of the People's Republic of China and the "barefoot" agricultural extension agents of the Green Revolution, who carried the message of new super high-yielding crops to farming populations in remote rural regions of the Third World. In many cases, the international technical and development assistance organizations have been the agents for transfer of this technology.

One theory underlying the concept of the barefoot microchip as a tool for development is that computer use may enable poorer nations to accelerate the development process. This acceleration involves "leap-frogging" over an intermediate period of industrial development and its economic infrastructure and organizational culture to the

Reprinted with permission from *Science and Public Policy*, the journal of the Science Policy Foundation, United Kingdom, February 1985, Volume 12, No. 1, pp. 15-20.

"Information Age," thus catching up economically with richer nations. Over the past two decades, there have been many proponents of this technological leap idea.[2] The international development assistance agencies and independent experts in development are not alone in spreading the concept. In Paris, the World Centre for Microelectronics and Human Resources was created in 1981 to monitor worldwide development of microelectronics in order to allow French industry to take advantage of it; the World Centre also was supposed to transfer microcomputer technology to TWCs.[3] At the corporate level, IBM has created the Africa Institute to begin using microcomputer technology in general, and the IBM PC in particular, as resources in development.[4] These are only a few of the more visible organizations promoting the barefoot microchip.

Computers have been used in TWCs since the 1960s,[5] and comparatively recently, the microcomputer has joined the larger mainframe and minicomputers in the technology transfer process. Because of improvement in memory technology (continual increases in storage density, speed of operation, and decrease in energy consumption), microcomputers can replace the larger minicomputers in many cases; they are smaller (some are portable), less expensive than minis, offer a wide range of potential software applications, are easy to use by people who have no technical background with computers, and are likely to meet the data-handling requirements of rural development projects. Microcomputers thus put a wide array of capabilities at the disposal of organizations and individuals involved in the development work.[6]

For these reasons, it is likely that microcomputer hardware and software will be increasingly used in computer technology transfer efforts to TWCs (excluding, of course, those countries, such as India, which have chosen their own strategy for self-reliant development and support of information technology).[7] Clearly, it would be advisable to try to determine a successful path for this transfer as early in the process as possible; what is needed are well-thought-out and well-defined transfer guidelines and policies. However, ensuring successful transfer[8] has received little or no attention within bilateral and multilateral agencies, and there are indications in some instances that the transfer process is already faltering, preventing or complicating effective use of the microcomputer.

In the transfer process, both in First and Third World countries, the computer is introduced to a particular information environment. Particularly in TWCs, that environment may consist of both informal information handling networks—word of mouth, personal memory—and formal ones—involving the use of tangible records.

We are concerned with the formal networks because they are jumping-off points toward computerization. In TWCs, these formal networks generally cannot be computerized without a great deal of preparation, sometimes over a period of years. For example, it is not unusual to see records, files, and papers stored unidentified in stacks in otherwise empty rooms in ministries and other official buildings, as well as in private businesses.

Pressures for Computerization

The history of technology transfer to TWCs, short as it is, abounds with failures and misconceived projects. The introduction of waterless toilets in Vietnam[9] and too-large tractors in Egypt[10] are cases in point. Nevertheless, pressures from many sources are combining to push TWCs rapidly and inexorably to adapt their information management systems to computer use. Thus, for example, the capacity of central ministries to generate timely and accurate current data and to use analytic, rather than descriptive, information in policy decision making is under pressure to develop. There are weak and missing links between research and decision making that might be strengthened by computerized data handling. Further pressures come from demands of international donor organizations for better and more timely accounting of their money and evaluations of project activities, as well as the applied marketing knowledge of the few multinational companies that dominate worldwide selling of computer technology.[11]

Increasing demand for information and communication technologies that are cost-effective, durable, and easy to use is also coming from within TWCs. More and more government ministries and specialized agencies are seeking to obtain their own computer systems for many kinds of applications.[12]

What follows is an example of a recent attempt at transfer that has so far not been successful. This project of the U.S. Agency for International Development (USAID) and the Central Tunisia Development Authority (CTDA) illustrates the results of an ill-conceived project to install microcomputers and train people to use them.

Central Tunisia Development Authority Project

The CTDA was created in 1979 to manage and promote rural development in an economically poor region of the country. A long-term, general USAID assistance program in Tunisia had as one of its objectives "to train CTDA staff in macro-social data management and evaluation,"[13] a process involving three steps to be completed from 1979 to 1981. First, the CTDA was to carry out an informant survey in the summer of 1979; second, two Apple II microcomputers were to be installed to process survey data; third, the results of the survey were to be used as a means to build up the data management skills of the CTDA staff. A USAID Area Development Subproject was created to carry out the work. The rationale, in part, was to evaluate the success of the subproject itself.

A three-member U.S. consulting team working with USAID went to Kasserine, the CTDA's headquarters, to install the microcomputers and run a seven-week training course for personnel to learn how to use the micros for handling the survey data. It was hoped that at the end of the course, the CTDA would be able to produce reports on semiannual updates of the survey by itself. Eventually, its payroll and other accounting tasks would also be put on the micros. (In 1979, the team had made a preliminary visit to the CTDA to familiarize itself with the working environment;[14] the team also made other visits of several weeks each to Kasserine both before and after the training course.)

The team planned the training course for the summer of 1980 to train staff and overcome any technical computer installation problems. It was to be a tightly scheduled period in which carefully arranged lessons based on the survey were to be used to train the operators. The team's goals were to set up the computers and peripheral equipment, train the staff to use the computers, analyze data output, and produce technical reports.

The team also had four of its own objectives—apart from those related to its AID project—in connection with an information system it had developed for use with microcomputers in development situations. Team members hoped to demonstrate that it is possible to install an autonomous on-site information handling and analysis system for regional units of government to do applied research on local needs and follow up the results; that microcomputers would allow this to be done cheaply; that no previous experience was necessary for local personnel to run the computers; and that the programs introduced were usable in a wide variety of applications.

Difficulties Encountered

The team began training CTDA staff to use the micros, and soon ran into some unexpected difficulties. It quickly became clear that the trainees were dissatisfied with the relevance and accuracy of the survey and proposed instead to use the hands-on computer training time to put their own projects on the computer. The team agreed to this; some students used agricultural data from a regional project and others used data from an irrigation project. The survey data were not computerized. To some extent, both the training and the entire project were affected by the fact that the CTDA, having been only very recently created, was preoccupied by its efforts to win the respect and cooperation of other regional and local agencies with which it had to coordinate its activities.

There were also unexpected technical problems to be solved in the installation of the computers, as well as some environmental hazards to the machines, such as excessive dust and cigarette smoke, so that they required almost constant attention by one team member. Other logistical problems related to the micros cropped up: the CTDA offices were located in five different buildings in 1980 and 1981, and one change of location involved moving the computers as well. Both micros were out of order for about a year, apparently because stabilizers were not used to neutralize the effects of power fluctuations in Kasserine. One computer was repaired; the other has been abandoned as irreparable. Finally, the use of the micros was not promoted at the CTDA during the periods between the consultants' visits because top CTDA management attached little importance to the regional/coordinating role of the CTDA, and also to data collection and analysis in general.

Despite all this, in the course of the first few weeks of training, the team did see progress and hoped that most of its originally planned activities could be accommodated in the remaining time. However, shortly before the end of the course, the CTDA director requested a report from the informant survey data. Because the director and the consultants had different ideas about its purpose, it is not clear whether the report was to be part of the CTDA's contribution to a new Five-Year Plan that was being prepared, or whether it was needed for some public relations purpose. In any case, the team was forced to suspend training in order to enter and analyze the survey data for the report. By the time the report was finished, the seven weeks had passed, and the team left Kasserine.

Reasons for Abandoning Survey

As of September 1984, the survey and updates have been completely abandoned for a number of reasons, including lack of personnel in the CTDA Planning and Evaluation Directorate (the team had sent a CTDA staff member to the United States to take a ten-week course in computer maintenance and operation; this man left the CTDA upon his return to Tunisia). Another reason may be a critical lack of interest at high levels in the CTDA for the duration of the AID subproject. Finally, the regional evaluation system based on periodic informant surveys was abandoned in favor of other methods of evaluating the subproject. The other task originally intended for computerization—the payroll—was not run until February 1984. It still has so many bugs in it that it must also be calculated manually each month.

By its own assessment,[15] the team left behind ill-trained operators and a computer system that remained inadequately installed, due only partly to an unexpected move of the CTDA offices shortly before the team left Tunisia. However, one of the team members concluded that, in terms of its own objectives, the seven-week effort had been a success.[16] This is puzzling because, despite a great deal of technical assistance and follow-up, in the three years since the team was in Tunisia, the hardware has not been successfully installed and is not performing tasks a local or regional unit of government might find useful. Although the initial cash outlay for the microcomputers, peripherals, and software was low, the cost since 1980 of trying to install the micro that is not yet functioning has become much higher.

Finally, it seems doubtful that the wide applicability of the programs introduced has been demonstrated, since in three years, not even the CTDA's monthly payroll is successfully computerized. However, it should be mentioned that the authority has had an agreement since early 1983 with a Tunisian management consulting firm to help the CTDA develop workable budget, payroll, and accounting programs.

After several follow-up visits, the team concluded that it had tried to do too much too fast and that it might have been better to concentrate only on the basics of system installation and training. As it was, its syllabus for the seven-week period was tightly filled with lesson plans and other activities that may have prohibited flexibility and concern with the actual needs of the CTDA staff. According to the team's report on their final trip to the CTDA in the summer of 1981: "In retrospect, all that should have been attempted was to deliver simple basic skills working shoulder to shoulder with the Tunisians and let it go at that."[17]

Belated Success: Egyptian Health Information System

A different approach to the transfer of the microcomputer is outlined in the following case study of the National Health Information System, a project of the government of Egypt.[18] In 1981, the Ministry of Health proposed the system as part of a reorientation of its approach to health problems by examining current health needs and establishing priorities for meeting these needs. To achieve its goal, it was deemed essential that a new information system be created. The system's objectives were to gather data at the local level on how current health programs were proceeding, such as immunization, fertility regulation, urinary schistosomiasis among school children, tuberculosis, and rheumatic fever control. Other areas of interest were the relative effectiveness of technical equipment, training needs, the cost and cost-effectiveness of various parts of each program, and an overall assessment of the efficiency of health interventions compared to the Ministry of Health's targets for each programme.

The data were to be gathered by training personnel in 3500 health facilities (district hospitals, rural centers, local health units, emergency care units, family planning centers) spread through 250 districts in Egypt. The amount of data thus collected would be too great to be worked with manually; therefore, it was decided to install a microcomputer in each district. In addition, electronic data processing would provide quick transfer of raw data into meaningful output, allowing timely action if necessary, such as in the discovery and eradication of an epidemic.

A detailed needs assessment[19] was done comparing various microcomputers and software written specially for the Egyptian system, allowing entry of data and printouts in Arabic. The study also examined the data processing workload required to satisfy the Health Ministry's needs. The raw data would be gathered monthly at the local facility by trained clerks recording answers to the same questions each month on simple forms.

Another source of data was the statistics department and the Bureau of Health in the ministry, containing figures on population distribution according to sex and age, birth distributed according to the mother's age and her medication supervision, as well as rank of the child among siblings, village population and number of families in each village in the district, and tuberculosis control. Finally, the study outlined the hardware and software requirements for processing all the data from these sources, and for producing monthly and yearly reports for the Ministry of Health.

Recording Of Data

Once it was decided, on the basis of the study, which hardware and software to use for the system, two microcomputers and programs were acquired for testing under actual environmental and manpower conditions. In the course of the testing, however, it was found that the data to be processed were incomplete, and what data did exist were not accurately or fully recorded on the forms designated for the project.

Testing and implementation of the information system was suspended until more basic issues of manual data handling were resolved.

This became a process first of thoroughly overhauling the data collection system at the local facility level. People were sent to work with the clerks responsible for data collection and demonstrate the need for accurate and complete recording of data. Second, it involved upgrading the statistical apparatus within the Health Ministry because the accuracy of its statistics was essential if the information system was to produce usable indicators of the efficacy of Egyptian health programs.

This process has taken two years, during which one microcomputer went unused and the other was used for routine word processing and clerical work in one of the Health Ministry's offices. The two years were not wasted time. Dr. Aziz El Kholy, the director of the medicine and health care sector of the National Information for Egyptian Development project,[20] summarizes the positive effect of the two-year period before actual testing and use of the microcomputers could begin. "The microcomputer, just by being there, became the motivation for people to change their thinking and information-gathering behavior," he says. "It forced them to become organized and also to solve the problems of effective and efficient data collection." He advises others concerned or involved in microcomputer transfers to ignore the details of hardware and software issues entirely, "fossilize one's ambitions" for the project, and be prepared to wait however long it takes to do the preliminary organizational work before concentrating on the technology. The question thus arises whether in related development projects, microcomputers ought to be purchased merely to be an incentive for better information management. This may not be as absurd as it sounds.

Microchip: Distraction from Development

"The microchip could well distract the attention of people from development," warned Gordon Temple of the World Bank at the United Nations Development Forum's Tenth Anniversary colloquium.[21] This is precisely what has happened in the case of microcomputer technology if we approach the process of development partly as one of organizing information into a system suitable for eventual computer use. The practically-abandoned USAID subproject in Tunisia illustrates this point. In the course of introducing computers to the CTDA, not enough attention was paid to problems presented by the existing information handling environment and, in particular, its differences from the environment in developed countries. Identifying and preparing for these differences is crucial to the effective introduction of a computer under any circumstances, regardless of relative state of "development." El Kholy[22] aptly characterizes the peculiar situation created by the transfer of microcomputer technology: "If you bring in a tractor or drilling rig all you need to do is train people to run that particular machine and maintain it. But successfully adapting the microcomputer in Egypt has nothing to do with simply teaching people how to run the machine. First you have to successfully run your office or laboratory or department before the computer can be of any help at all."

Thus, in the Egyptian health services case, one might say that the technology acted as a probe to reveal that basic tasks and procedures for data and file organization, statistical calculation, records management, program performance review, and report production had to be dealt with before the microcomputer could be used. In the two years since the microcomputers arrived in Cairo, the health organization has been getting its house in order so that the microcomputers can be used. It may have become more "developed" (in the sense of acquiring new skills and new attitudes toward a particular kind of organization and administration) during this housekeeping interval than when the computers are actually used. Thus, the process of development in the case of transfer of microcomputer technology could be said to be one of preparing for use of the computer.

Future: Limited Prognosis

The question arises why the microcomputer should be transferred to TWCs at all—countries with myriad problems of providing for the basic needs of their people and relatively few financial or technological resources to break the dependency link with First World nations. This is a question that cannot be addressed here. It is sufficient to point out that reductions in the costs of computer logic devices and computer memory (at the rates of 25 percent and 40 percent per annum, respectively) and technical developments in the man-machine interface that have made computers increasingly "friendly" and provided a wide array of applications are likely to make the transfer of the microcomputer to the developing world a *fait accompli*. Thus, it is extremely important to concentrate on the most effective and successful means of transfer and utilization as discussed in this chapter and elsewhere.[23]

Given the transfer of microcomputer technology at present and in the future, what might its potential be if and when the technology is successfully introduced? The microcomputer could conceivably become a significant, even revolutionary, tool in development efforts. In brief, using the microcomputer, provincial, district, and subdistrict levels of government could begin taking over many of the tasks central ministries with their microcomputers or mainframe computers cannot, or should not, perform.[24] The local entities could help coordinate activities and services provided by extention services, cooperatives, or village banks and report back to central ministries. These would then be in a position to compare the efforts, say in health services, or different districts and to get a more accurate and timely picture of activities in lower government levels.

Once more, it is appropriate to point out that microcomputers are becoming less and less expensive, with the "throw-away" $100 computer not too far off. Portable, battery-operated microcomputers could be used in field operations and remote communications; some microcomputers have been shown to be durable and efficient in hot, humid, and dusty rural environments. Their ruggedness, reliability, and range of capabilities make them excellent tools for information handling in the field, in small towns, and even in isolated villages; used as a field tool, the microcomputer would make possible data entry and checking at the survey site.[25]

Finally, there is a paradox worth mentioning in discussing the use of the microcomputer for development. The same qualities that make the microcomputer an attractive development tool also perversely contribute to its misuse. Precisely because it is so cheap, there is temptation to throw it at situations in which a computer might be desirable, but the circumstances prescribe caution. For example, a local government ministry, in the assessment of a development agency, may not be ready for a computer, yet may indicate its desire for a prestigious and impressive minicomputer. All too often in these cases, one or more microcomputers may be acquired for the ministry in order to gain time, or simply to be transferred for "diplomatic" reasons until circumstances are deemed more conducive to a successful transfer of a larger computer. One USAID officer in Egypt in charge of data management[26] says that the agency sometimes recommends microcomputers to Egyptian ministries because they are simpler and cheaper for "easing Egypt into computer use" than minicomputers or mainframes. This practice threatens to deflect attention from the administrative, organizational, and basic record-keeping problems that can mean failure in the transfer process.

As the microcomputer becomes a more widely used tool for development, it will no doubt have its share of failures and successes just as the transfer of any other new technology has, but the microcomputer holds a great deal of potential for indirect, and admittedly far-off, change in how information is handled, how it flows within organizations, and even in how people think about it in the TWCs. Therefore, it is perhaps not in vain to try to determine a successful path for it as early in the transfer process as possible.

Recommendations

It is not the purpose of this article to criticize development agencies or consulting firms, but to draw attention to a problem that does not seem to have received much consideration. In this spirit, the following suggestions are offered:

1. The transfer of microcomputer technology has been distinguished from the transfer of any other mechanized technology, for example, transportation and irrigation, particularly in terms of the amount of time necessary for effective transfer. The distinction implies that the transfer of microcomputers must be approached differently from other technology transfers. This means generally that the structure and administrative environment in each place in which micros are introduced will have direct bearing on the effective use of the technology. Development assistance agencies and organizations must formulate transfer policies and guidelines that make effective transfer possible under these conditions.

2. Too often, project planners are more interested in introducing the most recent hardware and software configurations than in the particular problems of the organizational structure and administrative and managerial systems in which the technology is to be housed. The main functions and applications of each proposed computer system should not be decided upon until someone, perhaps a data processing advisor, conducts a comprehensive analysis of the receiving organization's problems and needs. Most importantly, this assessment must include a thorough analysis of the organization's existing systems for manually organizing and managing all of the data that will be processed by computer. In this connection, such organizations might be better served initially by consultants in office management and organizational strategy rather than with consultants who recommend and install the hardware and software for computer use.

3. Specific procedural and policy recommendations based on such an analysis should be made to improve the existing data handling system (such as forms design, data collection, work flow, information management and retrieval). If they were put into practice, these recommendations would prepare the organization better for the transition to computer use.

4. A phenomenon that is especially prevalent in the transfer of computer technology is that of introducing too much of it too soon, usually because needs are assessed based on situations in developed countries. It is therefore important to begin by considering only the bare minimum of technology that is required to accommodate fundamental applications and functions for a particular project. More so-

phisticated technology can always be added gradually later. This simple, incremental approach has the advantage that better use can be made of the constant and rapid improvements in the technology.

5. It is important that training courses and software be oriented toward practical applications and easy hands-on use and experience. Course syllabi should be flexible enough to be simplified or abbreviated according to the specific training needs of the students. (For example, in training the CTDA personnel in Tunisia, the consultants might have prepared to teach the minimum of computer functions and applications necessary, and then been flexible enough to cut back even on these in order to ensure thorough training in whatever was taught.)

6. It is also important to present as much material as possible, including software, in the local language so that participation is not limited to those who speak and read English or French rather than, say, Arabic.

Acknowledgments

Research for this chapter was conducted with the support of the Education, Agroclimate, and Technology Program of Sierra Blanca Associates and Barefoot Computer Inc., both of Boulder, Colorado. The views put forward are those of the authors only.

Notes

1. From *Science and Public Policy*, Volume 12, No. 1, February 1985, with permission.

2. E. Berg, "The Barefoot Microchip: A Kind of Contradictor," *United Nations Development Forum*, March 1983, p. 7; R. A. Simmons, "Guidelines for the Use of Computer Technology in the Developing Countries," *National Bureau of Standards Interim Report No. NBSIR 73–423*, December 1973; C. W. Sprowls, "Can Developing Countries Take a Quantum Jump in Utilizing Computers?," *Proceedings of the International Conference on Computer Applications in Developing Countries*, (Bangkok, August 22–25, 1977), pp. 17–36; A. C. Clarke, "Third World 2001," *South*, November 1981, pp. 13–15; J.-J. Servan-Schrieber, *World Challenge: Organization of Petroleum Exporting Countries and the New Global Order*, (New York: Simon and Schuster, 1981).

3. Danielle Rouard, "Au Centre Mondial Informatique," *Le Monde Aujourd'hui*, April 15–16, 1984; J. T. Simonds, "Centre Mondial Informatique et Ressources Humaines International Actions," *UNIDO presentation*, March 20, 1984, available from the Centre Mondial Informatique et Ressources Humaines, 22, Avenue Matignon, 75008 Paris.

4. Personal communication, L. Lipp, "IBM Africa Middle East," Neuilly-sur-Seine, France, July 2, 1984.

5. S. Nilsen, "The Use of Computer Technology in Some Developing Countries," *International Social Science Journal*, 31 (3), 1975, pp. 513–528.

6. H. D. Toong and A. Gupta, "Personal Computers," *Scientific American*, 247 (6), 1982, pp. 86–107.

7. B. Nag, "Appropriate Computerization: An Instrument of Development and Social Change in Developing Countries," in J. M. Bennett and R. E. Kalman, (eds.), *Computers in Developing Nations: Proceedings of the IFIP International Seminar on Computers in Developing Nations (Melbourne, Australia, 13 October 1980)*, (Amsterdam, North-Holland, 1981), pp. 53–57.

8. As always, the criteria used in deciding whether a project is "successful" or not are difficult to establish to the satisfaction of everyone. For the purposes of the issues addressed here, however, success would be determined according to the degree to which existing information handling systems have been adapted for use on a computer. This agrees with a loose definition of "development" later in this chapter as it relates to the transfer of microcomputer technology.

9. J. McMichael, (ed.), *Health in the Third World—Studies from Vietnam*, Nottingham, Bertrand Russell Peace Foundation, 1976).

10. Y. Walley, O. El Kholei, M. Abbas, and E. O. Heady, *Strategy for Agricultural Development in the Eighties for the Arab Republic of Egypt*, International Development Series Report No. 9, Center for Agricultural and Rural Development, Ames, Iowa, 1983. J. Minerney and G. Donaldson, *The Consequences of Farm Tractors in Pakistan*, Staff Working Paper 210, World Bank, Washington, D.C., 1975. F. Stewart, *Technology and Underdevelopment*, (Westview Press: Boulder, Colorado, 1977).

11. M. M. Horowitz, J. Nellis, F. Young, *Mission to the Commissariat General au Development Regional (Tunisia), USAID, Institute for Development Anthropology, and Clark University, Report No. DAN-1135-a-00—1078-00*, September 1983; E. Baark, "Appropriate Information Technology: A Cross-Cultural Perspective," *UNESCO J. of Information Science, Librarianship and Archives Administration*, 4(4), October-December 1982, pp. 263-268.

12. S. Nilsen, "Use of Computer Technology in Developing Countries," *Economic and Political Weekly, Review of Management (Bombay)*, 12 (48), 1977, pp. 110–114; UN Industrial Development Organization, *Implications of Micro-Electronics for Developing Coun-*

tries: A Preliminary Overview of Issues, UNIDO No. IS.246, (New York, 1982).

13. Personal communication, D. Young, U.S. Special Mission for Economic and Technical Cooperation Tunis, May 28, 1984.

14. U.S. Agency for International Development, Central Tunisia Rural Development Authority, Project No. 664-0312, Washington, D.C., 1979.

15. F. Young, *First Trip Report Assessing the Seven Week Training Course*, July 11, 1980, USAID, Tunis.

16. F. Bertoli, *Proposal for Computer-Based CTDA Project Monitoring*, July 28, 1980, USAID, Tunis.

17. F. Young, *Trip Report on the Team's Final Trip to the CTDA*, July 6, 1981, USAID, Tunis.

18. A component of the National Health Information for Egyptian Development Project, The (Egyptian) Academy of Scientific Research and Technology. See USAID, *A Proposed National Information Policy of Egypt, USAID PASA NF/EGY-0016-7-77*, September 1981.

19. S. Saleeb, M. El Akabawy, M. Riad, S. Girges, and H. Naguib, *Task Force Report, Project of the Government of Egypt, National Health Information System No. EGY/81/Tol/N71/99*, Cairo, 1981.

20. Personal interview with Dr. Aziz El Kholy (January 12, 1984), Cairo. See also G. Walker, *Notes on Rapid Appraisal of the Utilization of Rural Health Care Services*, Paper presented at the Conference on Rapid Rural Appraisal, Institute of Development Studies, University of Sussex, Brighton, U.K., December, 1979.

21. In C. Raghavan, "The Barefoot Microchip: Treasure Chest or Pandora's Box," *United Nations Development Forum*, March 1983, p. 6.

22. Personal interview with Dr. Aziz El Kholy (January 12, 1984) Cairo. See also Board on Sciences and Technology for International Development, *Microcomputers for Developing Countries Planning Meeting Summary*, U.S. National Academy of Sciences, Washington, D.C., September 26-27, 1983, p. 13.

23. See also M. Muller and W. Rayfield, *Proceedings of the International Conference on Computer Applications in Developing Countries* (Bangkok, August 22-25, 1977), pp. 1-6; V. Jorssen, "Management Information Systems and Developing Countries," in J. M. Bennett and R. E. Kalman (eds.), *Computers in Developing Nations: Proceedings of the IFIP International Seminar on Computers in*

Developing Nations (Melbourne, Australia, 13 October 1980), (Amsterdam, North–Holland, 1981), pp. 267–271; N. Jequier, "Intelligence Requirements and Information Management for Developing Countries," in R. C. O'Brien, (ed.), *Information, Economics and Power: The North–South Dimension*, (London: Hodder and Stoughton, 1983), pp. 122–140.

24. M. Y. Hamdy, *Computer/Data Acquisition Systems*, USAID Agriculture Mechanization Project Paper No. 263-0031, October 1983.

25. N. Berge and C. North, *Field Use of Microcomputers in USAID–Funded Projects*, USAID/ST/MD Contract, September 1983, available from Thunder & Associates, Inc., 319 Cameron St., Alexandria, Virginia 22314.

26. Personal interview with Lenny Jameson, Office of Data Management, USAID (January 10, 1984), Cairo.

3

Applications of Microcomputers in Third World Organizations

Carl H. Gotsch

Introduction

One of the enduring delights/frustrations for those working in the field of microcomputer applications is the unpredictable response of non-users to the idea of computerization. Attitudes seem to parallel the zero-one binary logic of the machines. One group, upon being asked about computers, responds that electronic data processing is the answer to all of the organization's information and managerial problems, i.e., computers will do everything. At the other extreme is the group that sees microcomputer technology as just one more inappropriate fad perpetuated by the developed countries. In their view, computers will contribute little or nothing toward improving organizational effectiveness.

Common sense dictates that the truth lies, as usual, somewhere in between. The Minister or Director who envisages that all of the organization's data are going to be available at the touch of a button is likely to be disappointed. However, the planning officer who envisages a more flexible approach to the budget, the financial officer who is trying to develop an up-to-date expenditure control system, and the researcher who wants to run some simple regressions to establish optimum fertilizer levels, all can be helped to fulfill their objectives.

It would be unfair to characterize all those who oppose experimentation with microcomputers in their organizations as technological

---

Reprinted with permission from the *Advanced Technology Alert Systems Bulletin*, a publication prepared by the Center for Science and Technology for Development of the United States Secretariat, June 1986, Issue No. 3.

"Luddites" who feel threatened by change.[1] Many have had previous experiences with mainframe systems that local human and material infrastructure could not support and are only too aware that the promises of computer advocates about "ease of use" have often remained unfulfilled.

But lack of familiarity has the potential for creating less well grounded responses. For example, senior officials may fear that they will be seen as less knowledgeable about the affairs of the organization than the junior officers who run the computer system. In the extreme case, their concerns may include the fear that their functions will be replaced.

In a variety of field activities carried out over the past four years, the Project on Microcomputers in Development at Stanford University's Food Research Institute has encountered virtually the entire gamut of possible reactions. Out of this experience have come certain rules of thumb that are helpful in identifying the circumstances in which the application of computers is likely to be successful.

Three main considerations need to be addressed. First, it is necessary to ascertain that the problem really is one in which a computer would help. It seems extraordinary that this point needs to be stressed, but by far the largest number of failures in the diffusion of computer technology occur for the simple reason that the machine does not add much to the decision–making environment. It sits in the basement gathering dust for precisely the same reasons that the models and reports with which it is often surrounded have shared a similar fate.

Second, if computers are to be installed successfully, care must be taken in the selection of hardware and software. As anyone even peripherally connected with the industry is aware, the variety of configurations and packages that are currently available is mind boggling. It is a textbook axiom that the system designer must consider the needs of the user in selecting the appropriate technology, but in a situation in which the technology is changing every six months, that is sometimes easier said than done.

The third component of a successful diffusion scheme is an accurate assessment of the pool (current and potential) of human resources that will be available to implement and maintain the system. Demands on human resources have been altered dramatically by the microcomputer revolution and an organization need no longer maintain a stable of systems programmers to guard the computer room. However, the irresistible tendency of people "selling" microcomputers has often led to a misunderstanding of what can and cannot be done with different skill levels. One week of training in a database management system does not equip a novice to write an accounting package regardless of what the advertisements promise.

Three Microcomputing Environments

In our work with Third World organizations, we have come to recognize three different microcomputing environments within which the concerns indicated earlier can be applied. Although not watertight, they form a typology which adds further refinement to the identification of those situations in which a program is likely to be successful.

First, there is the **planning and budgeting field**. It is in this area that the so-called "electronic spreadsheets" have been very effective. Spreadsheets, which organize data by rows and columns in ways familiar to anyone who has used a columnar pad, are probably the most accessible type of microcomputer software.[2] User intuition developed in a more traditional paper and pen environment is readily transferable and provides a useful point of departure in organizing data.

The power of the spreadsheet comes from several sources. Most obvious is the ability to designate cells whose values will be calculated from formulas involving other cells. This simple device opens the way for the so-called "what-if" computations. Changes can be made in assumptions or empirically determined coefficients and the computer will recalculate all of the "derived" cells in a matter of seconds. The rapidity with which these changes can be made is critical in situations where negotiating teams are confronting one another, e.g., during International Monetary Fund or World Bank visits, or where a wide variety of sensitivity tests are to be conducted on a feasibility study.

In addition to the "what-if" possibilities, the power of the software is enhanced by the ability to copy data and formulas from one part of the spreadsheet to another. This replication capability is important because it drastically reduces the time required to develop the spreadsheet. While the ability to do recalculations would be valuable under any circumstances, insertion of similar data and formulas in each cell with the assistance of the computer to do the formatting increases the power of the packages significantly.

In all of the more recent spreadsheets, a third source of software power is the ability to implement "macros." Macros are simply crude computer programs that capture the sequence of key strokes required for complicated operations and store them under a single keystroke. They may actually be written out on the spreadsheet or saved as a sequence of keystrokes, depending upon the spreadsheet being used. The function of macros is not only to add to the productivity of the experienced user, but to customize spreadsheet applications. Such customized "templates" upgrade the productivity of less skilled individuals in two ways. The most obvious is to help overcome limitations that have their origin in the lack of familiarity with computers. However, "macros" can also encode skills of a theoretical nature. For example, the implementation of benefit–cost analysis for investment projects is full of conceptual pitfalls. Many of these can be avoided if the calculations that are required are already embedded in a spreadsheet template. Intermediate level technical skills requiring only a modest amount of supervision can be employed to undertake a much larger portion of the data collection and computation than would be the case if only the standard textbook guides were available.

The second major microcomputer application is found in the area of **administration and control**. Common examples of the problems encountered in this field are financial accounting, inventory management, personnel records, vehicle records, etc. The administrative or management field is usually characterized by the need to enter and manipulate large bodies of data for the purpose of preparing periodic reports on the status of the system. For this purpose, a variety of microcomputer database management systems have been developed. Some are simple and feature ease of use rather than versatility. They are suitable for keeping track of data where retrieval is the most important need for the user. The strength of the more versatile packages, however, lies in the embedded "languages" that permit customization of the operation. Programs written in dBase II, for example, or Knowledgeman require only clerical skills to produce the reports needed by the organization's management.

The power of these so–called "fourth generation" languages over the regular high–level languages such as BASIC, COBAL, FORTRAN, and others results from the fact that much of the data handling code is already taken care of by the commands that make up the database management system language itself. As a result, the productivity of the individuals responsible for the development of the specific application can be increased significantly. In addition, a lower level or skill is required to maintain the system and to make minor modifications to it.

The need for an accurate and timely control mechanism is obvious when one imagines the plight of individuals who have been asked to make policy and program decisions without accurate information about the current status of the system. It is difficult, for example, to manage a government food price and rationing system if the size and distribution of food stocks is unknown. Food may be ordered when it is not needed; alternatively, food may not be available when areas are hit by drought and other natural disasters. "Early warning" systems which try to provide advance warning about possible production shortfalls are of only modest help if they cannot be compared quickly and accurately with the availability of stocks, both actual and in shipment. Political caution in all such cases dictates that a higher level of stocks be held than would be necessary if an effective inventory management system existed.

A similar problem occurs in the financial area. As every finance officer knows, the rate of expenditures among projects may vary widely as the budget year progresses. Without timely information on which projects are overspending or underspending, there is little that the organization's management can do to maximize the returns to funds available in the current period. When financial accounts are months behind, as they often are unless an efficient financial control system has been implemented, the lack of information is a serious bottleneck to allocating money from projects that are behind schedule to projects that have shown an ability to spend money rapidly.

The third and best understood computing environment is one which goes under the broad heading of "**research.**" The research paradigm and the analytical methods used have been well developed and are part of the researcher's quantitative training. Anyone trained in the field of statistics, for example, will find little difficulty in using one of the numerous statistical packages available on microcomputers to carry out traditional inference tests. In part, this is because a number of microcomputer packages are offshoots of mainframe applications with which the advanced researcher may already be familiar. More importantly, the computational algorithms are already implemented, which means that the only real hurdle is in learning the details of the input/output routines of the package.

Sophisticated packages required for manipulating large data sets and doing nonlinear calculations are only beginning to appear on the market. As the hardware/software section of this chapter notes, however, the present limitations are temporary. Prototype machines currently exist that, in a few years, will make the current mainframe environment available on the researcher's desktop. At that time, more demanding integrated mainframe systems such as SPSS and SAS will become available in their entirety.

The most interesting aspect of microcomputer use in the research field is not the implementation of traditional statistical and optimization methods, but rather in the new approaches they permit to data collection and data processing. In both developed and developing countries, the traditional sequence in conducting field work has been to undertake a survey, tabulate the data, and publish the results.

Economies of scale in data collection, data entry, and data analysis are so great that a more iterative approach in which data are entered and validated as they are being collected has not been feasible. With the advent of microcomputers, it has become possible to take a much more interactive approach to data gathering. In some cases, small lap-sized computers have been used to record interview responses directly. A somewhat less ambitious methodology is one in which initial returns are evaluated while the survey is still under way and corrective measures are taken while enumerators are still in the field.

The attitudes and morale of researchers in both the natural and social sciences is greatly improved if the results of quantitative work can be processed and published in a timely way. Most researchers working in developing countries have had some advanced training and hence are acquainted with basic quantitative techniques. However, the lack of opportunity to pursue this training within the local environment has impeded empirical investigations. Instead of attempting to examine the real world quantitatively, research often has deteriorated into broad descriptions devoid of analytical rigor. In the absence of adequate computational facilities, it is not surprising that this type of research has become prominent. If researchers cannot get simple statistical tests done without lengthy waits, empirical work is likely to be slighted in favor of a more descriptive mode.

Hardware and Software

Of the three types of computing environments, planning and budgeting is perhaps the least demanding in terms of hardware and software. The sizes of most of the exercises are such that the now standard 16-bit configuration with 256K memory is more than adequate for all of the available spreadsheets. For that matter, even 8-bit machines such as the Apple IIe, Kaypro, and Morrow Design, contain sufficient memory for a wide variety of budget and planning applications.

The same might be said about storage devices. The standard double density floppy diskettes will hold all but the largest spreadsheet models. Thus, if spreadsheets are to be the major application, the inclusion of a hard disk in the hardware configuration is probably an unnecessary expense.

There are a variety of software packages that will handle the most frequently encountered planning and budgeting applications. Lotus 1-2-3, Supercalc, and Multiplan are among the most widely advertised but there are others that would serve the purpose equally well. Perhaps the most significant question that users will want to ask is whether the spreadsheet lends itself to customization. That is, can the senior planning and budgeting officials, once they have designed and implemented the system, create an applications package in which a substantial part of the routine calculations can be left to more junior staff? This facility has been developed in an advanced form in the Lotus products 1–2–3 and Symphony.

An important acquisition in the planning and budgeting environment is software for word processing. Plans, budget documents, project feasibility studies, and contracts invariably require revisions as a result of negotiations among the various actors involved. Perhaps more than any other area, planning and budgeting departments benefit from the capabilities of a word processor to revise, reformat, and print out new versions of departmental documents.

The hardware and software requirements of the administration and control environment are more demanding than those of planning and budgeting applications. First, there is the need to store large amounts of data. If the financial system of a Ministry is fairly extensive, for example, the use of a hard disk for data storage is virtually a necessity. Lack of such a mass storage device places a heavy burden on the management of a floppy disk system that may run into dozens of data diskettes. The possibilities are obvious for error and confusion when the system is being run by people with only modest computer skills.

Because of the variety of applications, hardware and software requirements in the research area are the most difficult to characterize. In the field of optimization, for example, every increase in computing power (internal memory and processing speed) is welcome. Each such advance brings microcomputers closer to mainframes in their ability to handle large models with integer constraints, nonlinear objective functions, etc. As it is, current hardware on the 16–bit machines is sufficient to handle modest linear programming exercises (125 rows by 300 columns) efficiently. In the case of IBM PCs or PC clones, the speed of performance can be enhanced at modest cost by adding a Intel 8087 numerical co–processor. In the field of econometrics and statistical analysis, such packages as Micro TSP running on similar equipment with 256K of memory can do all of the normal statistical calculations, including somewhat more esoteric techniques such as two–stage least squares.

Hard disk devices also have alleviated much of the data storage problem encountered in earlier microcomputer–based field survey work. These drives are available in a variety of sizes and units with 20, 50, 70 megabytes and above currently available on the market.

The rapidly declining prices of this type of storage device means that except for groups engaged in conducting national censuses or national sample surveys or that have large billing and accounting functions, the microcomputer environment is capable of handling most processing tasks at nominal cost.

Training and System Maintenance

In our work in the field, we have distinguished four skill levels that most organizations will require. (There are additional, higher levels that will be needed if complicated applications software is to be developed, but few groups will need to be concerned about maintaining this type of expertise in-house.) First, there is the simple data entry skill (Level I) that can be taught in a few hours. This type of expertise is more a function of the care with which the user approaches the work and typing ability than any interest in or knowledge about computers. Lack of this level of skill has not proved to be a serious bottleneck to the use of machines in developing countries.

Level II skills are somewhat more demanding in that they require the individual to be knowledgeable about the computer and its operating system. Level II people need to be able to install commercial packages, to back–up materials, to diagnose simple problems having to do with electrical connections, peripheral hook–ups, etc. They should be knowledgeable about the programs they are being asked to implement and to have some understanding of the software packages that underlie these programs. Most graduates from local universities can acquire these skills although, because of its greater rigor, a major in science or engineering is helpful.

In addition to Level II skills, Level III people should be able to see that production schedules are met, make minor modifications in the programs and procedures that have been set up, produce simple *ad hoc* reports from existing databases and, in general, supervise the day–to–day operations of several workstations.

Level IV skills are required for developing modest custom applications. Individuals should be able to diagnose most computer–related problems and determine what type of repair or maintenance measures should be undertaken. While modern database management languages do not make a background in computer programming mandatory, persons with some previous experience should be able to sit down with an organization's management and to do rudimentary design work that would result in simple management information systems. (Very complicated diagnosis and design requires higher levels of skill and experience than probably can be maintained within the organization.)

The mix of skill requirements varies by type of computing environment. In the field of planning and budgeting, the data entry part of the work is relatively modest. Consequently, the pyramid of skills has fewer low-level people and more intermediate- and high-level skills. This is because the development of a benefit-cost analysis or the creation of a budget requires skills beyond those involved in manipulating the machine. Even with the customization of calculations, a certain amount of familiarity with theory and methodology is needed.

In administration and control, the skill pyramid can be sharply hierarchical. *After* its implementation, a very few individuals with Level III and Level IV skills can supervise a substantial data processing operation. Once the reports that are needed have been agreed upon within the organization, the bulk of the work is in data entry, a process that requires only Level I clerical skills.

Because of the broad range of topics it covers, training and system maintenance in research is again difficult to characterize. In general, however, because of the development of more accessible packages, Level IV skills, i.e., those of a programmer and systems designer, are less in evidence around research establishments using microcomputers. The use of spreadsheets and database management packages tends to involve researchers personally in an interactive mode. There is relatively little need, therefore, to develop the skills needed to write the command files or procedures that would simplify repeated applications of the same data processing methods.

If the group does field work, of course, there may be a substantial data processing operation. It is somewhat different from the periodic work of an accounting and financial management department, but this type of work does have a heavy data entry component and hence may require a substantial number of people who have only Level I skills.

Institutionalization of Microcomputer Applications

In our experience, the most difficult part of realizing the possibilities of microcomputers in Third World organizations has to do with institutionalizing their use. The initial stages of introduction tend to be highly personalized and can be initiated with outside assistance. If the appropriate subject matter has been chosen, it is not hard to create a certain momentum that moves the organization into the area of electronic data processing.

However, the initial spurt of enthusiasm, important though it may be, cannot assure the long-run success of the venture. A number of substantive and administrative changes will need to be made within the organization to insure that the system will continue to grow and develop. At the most mundane (and therefore often neglected) level, someone has to be made responsible for the machines and their functioning. One of the great strengths of the microcomputer, namely its decentralization, becomes a problem when servicing and maintenance are required or additional supplies are needed; no one feels responsible. Some type of conscious administrative solution must be found if the machines are to maintain their productivity. (Note that this does *not* necessarily mean centralization!)

Perhaps more serious is the difficulty, in a decentralized microcomputer environment, of developing career paths for the individuals who specialize in using and supervising departmental machines. Unlike mainframe installations in which the creation of a computer center also leads easily to the designation of career paths for people who make the system work, scattering key individuals throughout a ministry or agency makes career identification more difficult.

With the advent of microcomputer technology, administrators and decision-makers in Third World organizations can have the benefits of electronic data processing at a fraction of the financial cost of installing a mainframe system. But there are other costs involved that are often unappreciated, e.g., training people and reorganizing the flow of information in their agencies. Failure to deal adequately with the institutionalization issue can result in systems that operate at far less than their technical capacity.

Notes

1. "Luddites" were an early nineteenth century group in England who protested the march of industrial technology by attempting, quite literally, to destroy machines.

2. In analyses of the history of the microcomputer industry, it is often remarked that VisiCalc made Apple Computer, not vice versa.

4

Energy Applications of Microcomputers in Developing Countries

Philip F. Palmedo

Introduction

Perhaps the most common adjective used to describe the rapid development and use of the computer is "revolutionary." Several authors liken the effects of computerization on societies and economies to those of the industrial revolution.[1] As Stehling has pointed out, this "second industrial revolution" is less obvious since it is characterized by miniaturization rather than largeness; by tiny silicon chips hidden inside devices and products rather than billowing smokestacks. More subtle, perhaps, than its predecessor, this revolution may be more profound, since it extends man's intelligence rather than his physical abilities.[2]

Those concerned with social and economic growth in developing countries must start to deal seriously with the implications of the computer revolution on those countries. Computerization is proceeding at an extraordinary pace and with remarkable pervasiveness throughout industrialized economies. If, as many expect, computers have significant economic value, they may add another critical possession to the "haves," distancing them even further from the "have not" nations.

While this broad issue is worth serious attention by the development community, it can be addressed, at least in part, by examining applications of computers to specific problems of development and development planning. In this chapter, we consider the applications of microcomputers to the development of the energy sector in developing countries and, in particular, in the process of planning and implementing national energy programs.

As we shall see, applications of microcomputers in the energy sector can be taken as representive of the more general issues of the role of computers in the development process. Particularly in this context, the term "computers" as used here should be interpreted broadly to include software, robotics, and database and information systems, what in French is referred to as "informatique."

Special attention, however, is focussed on the microcomputer. As we shall show through specific examples, the advent of the microcomputer has radically changed not only *how* analysis is carried out in the energy field, but *what* analysis is performed.

There are many ways of characterizing technical progress in computers: for example, the amount of information on a single silicon chip, circuit complexity per unit, and cost per unit operation per second. Whatever criterion is used, however, technical progress has been exponential over the last decade, with extraordinary rates of change from 20 to 30 percent per year.[3] There is every expectation that such growth in capability will continue for some time into the future.[4] Accordingly, a question which often arises when considering the application of microcomputers to energy planning in developing countries is whether next year's technical development will make this year's application obsolete.

There are basically two issues here. The first issue is simply one of capability. Is the current generation of microcomputers adequate in terms of memory size and speed for energy applications and, in particular, for energy planning? Although one could certainly find examples of programs used for energy analysis in industrialized countries that would not fit on a microcomputer, one can also argue that such models (which may involve extensive time series in highly disaggregated regions and the resultant manipulation of large matrices) are inappropriate for the vast majority of developing countries. Neither the quality of existing data nor the immediate planning needs warrant such modeling. In our judgment, and this is confirmed by a consensus of those working in this area, current microcomputers have quite adequate capability for the kinds of applications discussed in the chapters to follow.

What about cost? The important point here is that the cost of a microcomputer is small (usually much less than 10 percent, often less than 1 percent) compared to the total cost of an initial national energy planning effort, particularly if that effort involves foreign consultants.

The costs of those planning efforts (in national energy, the electric sector, or industrial conservation) are themselves small compared to the investments which they are intended to guide.

Applications

The range of potential applications of microcomputers in developing countries for energy development is extremely large. Since the objective of this chapter is not to catalog all possible uses but rather to arrive at useful policy conclusions, we have elected to examine some representative applications of microcomputers: in planning, energy conservation, and training.

Active interest in national energy planning in developing countries dates from the time of the 1973–74 oil price increases. Some of the earliest major efforts were carried out in Egypt and Peru under U.S. Department of Energy sponsorship. In the decade since that time, a large number of developing countries have instituted national energy planning efforts. Within the last four years, it has become common for these to be based on the use of microcomputers.

A wide variety of analytical activities fall under the general heading of energy planning:

1. *Energy Information Systems*: embrace two distinct concepts; first, development of a national energy database; second, development and operations of a system for collecting, storing, and sharing energy planning information.
2. *National Energy Surveys*: provision for collecting, storing, and sharing information that is not normally the responsibility of an existing organization.
3. *Energy Supply and Demand Analysis*: characterization of the national energy system from resource production through final consumption. Presents information by fuel type, by use/process, by demand sector. May also be regionalized.
4. *Analysis and Evaluation*: uses the information available to define energy supply and demand trends, energy/economic relationships, substitution possibilities, and the like.
5. *Economic and Pricing Studies*: analyses of the effects of energy consumption on the balance of payments, as well as evaluations of the effects of alternative energy prices.
6. *National Projections and Simulations*: projections of supply–demand balances with qualitative information to assess future energy supply and demand patterns, and the impacts thereof.
7. *Sector Specific Surveys, Models, and Studies*: disaggregates national data to present detailed sectoral and subsectoral resources,

technology, supply and demand characterizations, and projections.

8. *Energy Conservation Analysis*: uses previous national work and generic operating and technology information to evaluate the national opportunities for energy and economic savings through increased process efficiency.
9. *Macro–Economic Studies*: evaluates the effects of global market changes on the national energy and economic systems.
10. *Portfolio and Optimization Analysis*: provide for the selection of one or more optimal projects from among a portfolio of projects given a specific set of financial and other criteria. Also selection of optional supply and demand patterns given specific system characteristics.
11. *National Investment and Financial Planning*: allows for the selection of a mix of development projects given national cash flow, balance of payment, and other criteria.
12. *Input–Output Models*: relate structural shifts in national and regional economic activities to changes in energy supply, consumption, and price (and vice–versa).

For each of the activities listed above one can find an instance of the use of microcomputers in one or more countries. In integrated energy analysis most of the activities would be involved. There are very few examples, however, of such extensive analysis – with or without computers – in developing countries. To illustrate the use of microcomputers in various aspects of energy planning, in Table 4.1 we cite some specific recent applications in a variety of countries.

After several years of applications of microcomputers in planning, some general conclusions can be drawn. The development of microcomputer technology itself has facilitated both the process of policy analysis and the process of local participation. Thus, in several countries, including the Dominican Republic, Morocco, Tunisia, Sri Lanka, Indonesia, and Liberia, a more participatory approach to computer–based analysis is beginning to flourish.

The results of such cooperative efforts thus far are encouraging. As national staff have more direct input and participation in the analytic efforts, the results are achieving greater support. The speed of the computers allow national staff to play with variations in input, thus allowing development of some idea of the "sensitivity" of analytic results to small variations in assumptions. The use of integrated planning models, even the relatively crude models now being used, have allowed evaluations and analyses that might not have been done under other circumstances.

TABLE 4.1
National energy models

Structural	Mainframe Models	Microcomputer Models	Where Used
Energy Data Bases	Energy Modeling Data Base (EMDB), Brookhaven National Laboratory; Energy Supply Planning Model, Bechtel Corp.; Energy Balance Data Base (EBDB), Argonne National Laboratory (ANL)	no comparable data base structures available	Tunisia, Sudan, Argentina, Egypt, Peru, and the DR
Energy Supply and Demand Balances	(1) Argonne Energy Model, ANL (2) Reference Energy System, BNL (3) LEAP, ESRG/Beijer (Kenya)	(1) Argonne Energy Model, ANL (2) Energy Systems Planning Model, DSI (under development) (3) RESGEN; IDEA (4) Enerstat (under development) (5) LEAP (being adapted) (6) RMA model	(1) RES used in Tunisia and Sudan (2) Morocco (4) no data proprietary (5) Kenya (6) Tunisia, Indonesia
Economic and Pricing Studies	various	(1) E/DI Energy Pricing Analysis Model (EEPAM), E/DI (2) Energy Finance Assessment Model (EFAM), IDEA	(1) Somalia and Liberia (2) Tunisia
National Energy/ Economic Analysis	(1) Midterm Energy Forecasting Systems (MEFS), DOE (2) A Macro Economic Model of Inflation and Growth in Korea, Korea Development Inst. (3) BEEAM, BNL	no comparable micromodels	none Korea Portugal, Korea
Energy Conservation			various models and data bases Tunisia, Togo Analysis & Models

TABLE 4.1 (cont.)

Structural	Mainframe Models	Microcomputer Models	Where Used
Sector Specific	National Coal Model; On-shore Oil and Gas Supply Model; Aggregate Refinery Model; Solar Heating and Cooling Model; Residential Energy Use Model; OR Industrial Model (ORNL); Electric Power System Development Model; Technology Assessment Model (SRI); Oil Market Simulation Model	no comparable public micro models (1) Fuelwood/Forestry Model, The Futures Group (2) various proprietary oil and gas systems models	none Several African countries no data
Optimization Models	(1) Brookhaven Energy System Optimization Model, BNL; (2) Regional Industrial Supply Model, DRI; (3) Project Independence Evaluation System, DOE (4) Systems Europe Optimization Model		(1) Energy Management Training Program, Dominican Republic (4) Thailand
National Investment and Financing		(1) Energy System Planning Model, DSI; (2) Finance and Cash Flow Analysis Model, IDEA; (3) Energy Investment Assessment Model, E/DI	(1) Morocco (3) Dominican Republic
Input/Output Models and Macroeconomic Models	(1) Energy Disaggregated Input–Output Model; (2) Nine-Sector Generalized Equilibrium Model; (3) Wharton Annual Industry Forecasting Model	no comparable models	none

Two examples of U.S. Agency for International Development (USAID) supported national planning are illustrative of those conclusions. In the first, Energy/Development International (E/DI) carried out a National Investment Planning Project for the Comision Nacional de Politica Energetica in the Dominican Republic. This work was novel in that it sought to compare energy supply projects and conservation efforts to relate macroeconomic and energy investment considerations to energy prices. Thus, it transcended general national energy planning which normally proceeds only to the level of broad national energy strategies.

The preliminary results of the initial phase were unexpected in that they generally supported coal import and final-fuel use projects over development of indigenous energy resources such as hydroelectric power.[5] A more general set of conclusions was that this type of work is tractable from an analytic point of view, that national staff could participate effectively in the planning and analysis process, and that microcomputers are tools of sufficient capability and sophistication to carry out complex analyses.

In the second example, Development Sciences Inc. (DSI) developed a software package to enhance the planning capabilities of the Ministry of Energy and Mines (MEM) in Morocco.[6] The system developed (Figure 4.1) links energy supply, demand, and other baseline conditions—the national energy system—to planners' expectations about the future—the scenario—and an inventory of development projects—the investment portfolio.

The model then determines the performance of the portfolio over a 20-year period in terms of energy demand/supply balances, and its financial and manpower resource requirements. With this information, the planner can then return to the beginning of the planning cycle and redefine the portfolio, change project characteristics, or modify alternatives such as prices or imports. Thus, an acceptable plan emerges as the result of several iterations, the preferred plan being chosen with an understanding of the consequences of alternative investment strategies.

The results of the project were strongly supportive of the application of microcomputers as planning tools. For example Ernst of DSI writes that:

> ...The Moroccan decision makers feel very comfortable with the way in which the information on possible decisions is organized and presented. ...Key decision makers in the Ministry of Energy and Mines suggest very strongly that this software package represents the kind of planning tool they *can and will use*, and ...The combination of interactive use

FIGURE 4.1
Structure of the Analytical Planning System

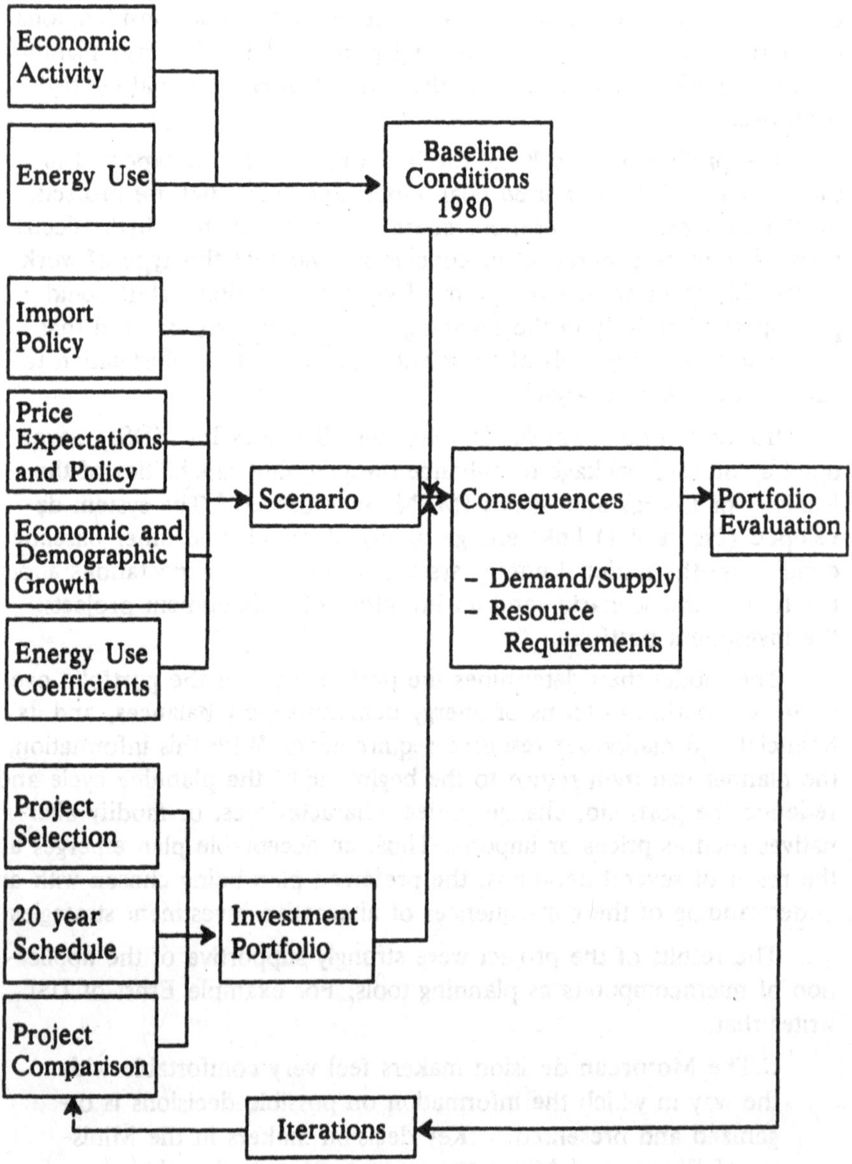

of the programs and the associated use of graphics also seemed to *enhance existing skills*.[7]

Industrial Energy Management

The use of computers for calculations in industry has developed over a period of many years. The oil and petrochemicals industries, for example, use linear programming techniques for planning purposes; most large oil refineries are operated with the help of a linear program model which optimizes the processing of a variety of feedstocks and the blending of intermediates into a wide range of products.

There are many possible applications for microcomputers in industry in developing countries. We would distinguish here two broad ones: analysis and process control.

An example of data management and analysis is the collection and analysis of plant-level energy data carried out by E/DI Europe in Tunisia.[8] Visits were made to 60 factories, and a questionnaire was completed for each by the survey engineers. Data collected included energy consumption figures for each energy form and corresponding production figures. The analyses carried out using these data and Versaform, an "electronic spread sheet," included a review of consumption for different subsectors and comparisons of energy efficiency with international data for corresponding manufacturing activities. Using the same software, it was possible to prepare a number of useful reports by searching the data files in different ways.

At the plant level, there are many calculations which are performed regularly by energy managers and engineers. Use of microcomputers allows those calculations to be carried out efficiently and to be part of more global, plant level analysis.

The microcomputer is being used extensively for energy planning and detailed energy conservation work in Sri Lanka. For example, Haggler-Bailly has carried out a variety of analyses of the textile industry using IBM-PCs. The analysis is used to propose an energy conservation program consisting of a number of options within given textile plants. Most important, the procedure reports monthly cash flow and other financial data required to obtain commercial bank loans for the project.

Of possibly greater significance is the application of microcomputers or microprocessors to process control in industry. Such applications can have significant impacts on overall plant efficiency and can "stretch" the supply of technically trained manpower.

Other Applications

There are many other current and potential applications of microcomputer in energy in developing countries. Some of them are:

- electric sector planning
- power equipment control
- refinery planning and operation
- renewable resource evaluations, (hydropower, wood supply–demand)
- building energy use analysis
- building HVAC control
- engineering design of equipment

Several of these are reviewed briefly in the report on the Sri Lanka Symposium.[9]

Training

One must distinguish first of all between training in the use of microcomputers and the use of microcomputers for training. We here refer particularly to the latter, as applied to energy.

The ability of the microcomputer to simulate energy supply–demand, energy flows in a factory, or heat balances in a building make it an extremely valuable tool for training in energy. For example, in a series of seminars held in Africa by Acres International and Lavalin (France), microcomputers were used extensively for training in conservation planning.[10] The IBM–PCs provided relief from the slowness and tedium of "number crunching," so the participants could focus better on the issues. The use of microcomputers in this instance also provided extensive "hands–on" analysis of energy conservation opportunities with real data.

Microcomputers are now used extensively to teach energy planning and analysis techniques in the Energy Management Training Program at the State University of New York at Stony Brook.

Combinations of microcomputers and video have proven to be very effective in the United States for technician training. In developing countries, such applications could be even more important, simply because of the serious lack of technically trained personnel.

Observations

Several observations and conclusions emerge from the earlier cited examples and from experience over the last several years of using microcomputers in developing countries.

Benefits

To a certain extent, direct benefits of computerization are predictable, often calculable, and have been observed in most cases where microcomputers, training, and support have been provided. Among these are the following:

- increased speed and decreased labor in repetitive, routine tasks involving the processing of large volumes of information;
- quality improvements in routine and nonroutine tasks involving processing and analyzing large volumes of information;
- integration and facilitation of *all* tasks involved in report preparation (data analysis, graphics and table development, word-processing), and improvements in the presentation of reports;
- increased data processing, storage, and exchange capabilities (particularly important in countries where such tasks are done manually);
- improvements in staff training and upgrading; and,
- extended analytic capabilities.

The indirect benefits of computerized energy planning at any technological scale are often difficult to characterize and vary from one setting to another. (They would, for example, not be so extensive in agencies that already use computers for other purposes as in agencies that do not currently use such tools.) Some tentative conclusions about these benefits follow.

1. The application of microcomputers in developing countries has the potential of narrowing the gap between decision makers and analysts. For various reasons, that process has the potential of significantly improving both the analysis and the decisions based on analysis.

2. The use of centralized computing facilities serving several agencies causes interorganizational friction resulting from disagreements about the costs of central services; physical access to the facility; distribution of computer time; distribution of remote access terminals; allocation of technical services; the potential relevation of proprietary information; and the like. Microcomputers contribute to easing such friction by providing a relatively inexpensive, highly capable tool for conducting work autonomously.

3. The use of microcomputers as analytic tools allows decisions to be based on more extensive and intensive criteria and to be made more expeditiously. It is to be expected that such decisions will be "better" (on average) than those of the past.

4. If, as many expect, computers have economic value, their use is of critical importance to developing countries, and would represent a quantum expansion in existing planning capabilities. If such tools are not used, planning capabilities will lag further behind developed countries.

In this regard, an important consideration is improving the distribution of competent talent in national planning agencies in developing countries. In many cases, there are only a few middle– to senior–level staff in each organization with good or better planning skills. Providing microcomputers will allow these officials to do more, better, and at the same time, assist in upgrading other staff.

5. There is one overriding benefit from the use of microcomputers for energy planning that is entirely distinct from items 1–4. That is, national energy planning organizations in developing countries can adapt existing planning models. By doing so, they can almost immediately apply the lessons learned over the past 10 years in developed countries without repeating the most common mistakes or incurring major development costs. Such acquisitions would expand their planning capabilities tremendously. There are attendent dangers to this adaptation, discussed in the following section.

Hardware/Software/Training

A key issue to be understood in judging the role of microcomputers in energy in developing countries is the balance between the following:

- Hardware
- Software
- Adaptation and application, and
- Training.

Traditionally, countries or firms have entered the computer age by deciding first on what computer hardware they needed or could afford. The next step in importance was the choice of software, which might have to undergo some slight modification. Financial and/or technical assistance was often required for hardware and software costs. More often than not, computer analysis of a major development problem was carried out by consultants in their own country. Thus, the traditional view and practice ascribed relative importance as follows:

The Traditional View:

Importance

Hardware Adaptation and Application **Software** **Training**

We would maintain that relative importance should be ascribed in exactly the opposite order. It is becoming widely recognized that software costs are growing while hardware costs are decreasing. More important, however, is the fact that existing models or software packages, for example, in the energy planning field are relatively useless by themselves. They must be adapted to the local situation, data must be collected for them, and often their structure must be modified to address new issues of importance. Furthermore, computer–based models provide only a partial set of tools to address important issues. Learning the limitations of models is often more important than learning to manipulate them.

In a very real sense, the ease of use of microcomputers brings with it a significant danger. The advent of the microcomputer, aimed as it was to mass audiences, brought with it the concept and the need for "user friendliness." For better or worse, "user friendliness" has become the norm, even in programs designed for the professional engineer or analyst. While admirable from most points of view, this phenomenon carries with it the danger that the analyst will have only a superficial relationship with the program he/she is using. One need know very little to be able to respond to multiple–choice questions asked by the computer. In some programs, the analyst need only insert a few numbers, choose a few options, and an analysis will be carried out, the results shown on fancy tables and graphs. This danger is particularly acute where specialized commercial software is being used or when a project is being carried out by foreign experts "in collaboration with" local technical staff.

Thus, the training component of a computerized planning project is far more complex than simply learning to operate the computer and run prepackaged software. It should include developing skills to modify existing programs and create new programs to meet new requirements. But, even beyond that, it should also include training in policy analysis as that analysis can be enhanced by computerization. Thus, for many applications of microcomputers to energy planning in developing countries, the training component is paramount.

Thus, we believe that the correct view of the relative importance of the components of a computer based planning program is as follows:

The Correct View:

Importance

Training Adaptation and Application Software Hardware

Larger Implications

This chapter has examined a few very specific applications of microcomputer technology in energy. The analysis is convincing—at least to the author—that: (1) microcomputers have distinct advantages over centralized computer programs in a developing country environment, and (2) there are several potentially beneficial applications of microcomputers in the planning and implementation of national energy programs. The advantages of computers in increasing the effectiveness of planning; in producing a more efficient system (e.g., through applications in the industrial sector); and in enhancing the transfer of technical expertise are clearly not limited to the energy field. This analysis confirms the hypothesis stated in the introduction to this chapter: computer technology can be of significant benefit to developing countries. Given the rate at which computers are being adopted by the industrialized world, the developing countries risk a further comparative disadvantage if they do not adopt them for their own social and developmental purposes.

There may be more to be gained than simple parity in applications. As we saw in the energy area, the appropriate use of microcomputers can enhance the transfer of technological information and expertise. Thus, in some instances, the computer "revolution" could be more advantageous to developing countries than to the industrialized ones. Modern computer technology can, in effect, be used to accelerate the process of technological development by short-cutting some of the more laborious historical transfer mechanisms. Although an example of this in industrial energy conservation has been discussed here, a more telling paradigm is perhaps medical diagnostic programs on battery operated micros. Such "expert systems" have the ability to upgrade the medical care provided by very large numbers of rural health workers.[11]

One can also imagine ways in which computers will work to the disadvantage of developing countries. Robots, for example, tend to replace repetitive low-skilled labor of the kind now sought out by industry in developing countries. When the hourly cost of robotic labor becomes sufficiently low, the competitive advantage of many developing exports will disappear.

Opposition to computers in developing countries is more apt to be based simply on argument of direct impact on employment.[12] There is often a resistance to the introduction of computers for just this reason. Such a perception was expressed not long ago by President Moi of Kenya. Although this concern contains a grain of truth, we feel that this is a short-term view detrimental to long-term national interests.

But, given this kind of perception, if developing countries are to profit from the long-term benefits of computers, they should be introduced with caution and into areas in which there are clear and immediate benefits. For example, in many countries there is a small but well-entrenched group of senior clerks who are now responsible for data compilation, accounting, and some analysis. This group could well feel threatened by the introduction of computers. However, if computers are introduced in an appropriate way, they can be used to upgrade the skills and prestige of that group, benefiting them substantially and ensuring their support of the change.

Introducing microcomputers first in fields such as energy and health can also be advantageous. In the energy field, there is a receptive and technically qualified cadre to adopt the technology and make it their own. The application of microcomputers to aspects of energy planning and management discussed in this chapter can have far-reaching implications for the development process.

In our introduction, we referred to the danger that microcomputers could benefit differentially the rich nations to the detriment of the poor. There is a parallel danger that, in the developing countries themselves, microcomputer technology will add to the welfare gap between the rich and the poor. This constitutes a further argument for a concerted effort to seek beneficial applications of microcomputers in areas of social benefit such as health, education, and energy supply. In all those cases, however, one must make a special effort to seek applications which bring benefits to lower income groups.

Acknowledgment

This chapter draws on a report prepared by Energy/Development International for the U.S. Agency for International Development entitled, "The Use of Microcomputers in Energy Planning for Developing Countries." The author has also benefited from his participation in the U.S. National Academy of Sciences (NAS) Panel on Microcomputers for Developing Countries and, in particular, from the discussions at the Symposium on the Use of Microcomputers for Management and Science held in November 1984 in Colombo, Sri Lanka, under the sponsorship of the NAS and the Computer Information and Technology Council of Sri Lanka. A paper prepared for that symposium by T. Owen Carroll entitled "Microcomputer Applications in Energy Sector Development" was of particular relevance to this paper.

Notes

1. For example, Philip H. Abelson, "We are in the early phases of shifts in the economy and in social patterns comparable in magni-

tude to those of the industrial revolution"; "The Revolution in Computers and Electronics," *Science*, 215, 751, February 12, 1982. Norbert Weiner was one of the first to recognize the revolutionary potential of computers, particularly as applied to the automation of factories. See Norbert Weiner, *The Human Use of Human Beings*, (Boston: Houghton Mifflin, 1950), p. 204, 205.

2. Kurt R. Stehling, *Computers and You* (World Publishing, 1972).

3. Lewis M. Branscomb, "Electronics and Computers: An Overview." Science *215*, 755, February 12, 1982.

4. Robert N. Nyoce, "Hardware Prospects and Limitations," in Michael L. Dertouzoo and Joel Mores, (eds.), *The Computer Age: A Twenty-Year View*, (Cambridge, Mass: MIT Press, 1981).

5. "Planning for National Energy Investment in the Dominican Republic: Phase I Report," Energy/Development International Division, IE/DI, Setauket, N.Y., March 1982.

6. From *Energy Planning Systems with Applications*, a section of DSI's Final Report to USAID regarding the work done in Morocco. This section was supplied by Ulrich F.W. Ernst, Development Sciences, Inc., Sagamore, Mass.

7. Letter from Ulrich F.W. Ernst (DSI), November 15, 1982.

8. E/DI Europe, "Industrialized Energy Study for Tunisia: Draft Final Report," prepared by E/DI Europe, London, U.K., for USAID (ST/EY), September 1982.

9. Mohan Munasinghe, Michael Dow and Jack Fritz, *Microcomputers for Development: Issues and Policy*, Proceedings of the First International Symposium on Microcomputer Applications in Developing Countries, a CINTEC-NAS Publication, Department of Government Printing, Sri Lanka, 1985.

10. J. Gillings, *et al.*, "Designing and Implementing National Energy Conservation Programs." Acres International/CDA Case Study. May 1984.

11. Expert systems in medicine have been under development for some time; see, for example P. H. Winston, *Artificial Intelligence*, (Addison-Wesley, 1977) and, more specifically, R. A. Miller, *et al.*, "Internist-I. An Experimental Computer-Based Diagnostic Consultant for General Internal Medicine," *New England Journal of Medicine*, *307*, 468, August 1982. Applications to primary health care in developing countries are being researched at the Centre Mondial d'Informatique et Ressources Humaines in Paris.

12. The issue of computers and employment in developing countries is highly complex and very much unsettled. See, for example, O. Pastre, "Informatisation et Emploi: des Mythes a la Realite," *Culture Technique*, No. 7, Paris, March 1982, and P. Stoneman, "New Technology and Employment: Extensions to the Case of Developing Countries," in F.G. Foster, (ed.), *Informatics and Industrial Development*, (Dublin: Tycooly International Publishing Ltd., 1982).

5

Kenya's Introduction of Microcomputers to Improve Budgeting and Financial Management in the Ministry of Agriculture

Thomas C. Pinckney
John M. Cohen
David K. Leonard

Introduction

The controversial Berg Report to the World Bank emphasizes the importance of inappropriate domestic policies in constraining development in sub-Saharan Africa.[1] The lack of effective policies for overcoming administrative constraints is given particular emphasis. While there is a rapidly increasing literature focused on improving the administration of agricultural development, financial and budget constraints are rarely mentioned.[2] The extensive literature on rural development in Kenya contains little or no analysis of financial management issues, even when focusing on the Ministry of Agriculture.[3] Similarly, the few references to budgeting in the African literature rarely get beyond generalities.[4]

An earlier version first appeared as Development Discussion Paper No. 169 of the Harvard Institute for International Development, Cambridge, Massachusetts, June 1984, "Kenya's Use of Microcomputers to Improve Budgeting and Financial Management in an Operating Ministry: An Updated Report."

By 1981, Kenyan officials, donors, and technical assistance personnel all realized that poor financial management was a major constraint to agricultural development in Kenya. At the same time, U.S. Agency for International Development (USAID) funded a series of demonstrations on the possible uses of microcomputers in the Ministry of Agriculture (MOA).[5] As a result of the demonstrations, the Ministry's financial managers decided to experiment with computerizing their expenditure monitoring system.

This experiment was one of the first to apply microcomputers to management problems in a development ministry. The early successes led to a series of publications analyzing the project as of mid-1982.[6] Since that time, much more has been learned about introducing and institutionalizing the use of microcomputers for financial management in development ministries. These additional lessons and the widespread interest in the project are the basis for this chapter.[7]

The background is set first by describing financial management in the Ministry in late 1981, with particular emphasis on budgeting, allocation of spending authority to field stations, and expenditure monitoring. Secondly, the reforms which have taken place using the microcomputers are described.[8] Thirdly, general observations are made concerning lessons learned about the use of microcomputers for financial management in less-developed countries.

Finally, it is important to describe briefly the history of the project which introduced the technology reviewed here. The management advisers referred to in this chapter are members of the Technical Assistance Pool (TAP), a multilaterally funded project begun in 1977, executed by the Harvard Institute for International Development, and aimed at assisting Kenya's Ministry of Agriculture to improve its agricultural planning activity and management systems. Between 1977 and 1980, the TAP averaged 8 to 10 advisers, most of whom were economists working in the Ministry's Development Planning Division. During this time, the two management advisers established a Management Systems Unit focused on administrative problems related to budgets, parastatals, personnel, vehicles, and so on. However, little progress was made until the microcomputer successes described herein. These increased Kenyan appreciation of what the Management Systems Unit could do, led to World Bank funding of two additional TAP advisers focused on financial and budget administration, and resulted in expanded adviser activities in assisting Kenyan officers committed to improved management of the Ministry.[9]

Situation in 1981

Overview of Budgetary Situation

An examination of Government of Kenya budgets since 1974 suggests two points. First, the coffee boom of 1976–78 caused budgets to increase much more rapidly than anticipated, the expansion continuing somewhat past the years of high coffee prices. Lower coffee and tea prices and high oil prices after 1979 forced the budget to expand less rapidly than anticipated in the 1973–83 plan. Secondly, the budget for the agricultural sector was cut more deeply compared to the plan than the government budget as a whole.[10]

This was seen as a result of the inability of the Ministry of Agriculture to spend its funds, particularly for development projects.[11] At the same time that the budget as a whole was being underspent, some donor–supported projects were having difficulties because of the absence of pledged Kenya Government contributions. The Integrated Agricultural Development Project of the World Bank was having severe accounting problems, and the Ministry had spent millions of dollars on foreign–aided projects for which no claims had been submitted to the donors. Clearly, MOA was having problems with its budget and financial management.[12]

The Budgetary Process

Kenya has a well–thought–out and carefully defined budget process, most of which is derived from its British colonial heritage.[13] The fiscal year runs from July to June. The budget process begins a year in advance, with each ministry preparing a three to five year forward budget, which is reviewed in discussions with the Treasury. This phase is intended to identify programs and projects, and establish priorities for funding. In December, the Treasury releases ceilings for the regular budget (known as the Estimates). At about the same time, field officers with spending authority submit estimates of their future fiscal requirements to the chiefs of the various program divisions in ministry headquarters. After modification, each chief submits a divisional estimate to the ministry's senior financial officer, who holds hearings with the chiefs and cuts excessive requests so that the total is close to the Treasury ceiling. The estimates are then submitted to the Treasury, along with a written defense for items with substantial increases compared to the previous year's budget. A Budget Supply Officer in the Treasury holds hearings, and either cuts the request down to the ceiling or builds a case to take to his superiors if over–ceiling expenditures seem justified. The Treasury holds final internal meetings to shape the budget for submission to Parliament. The Estimates are made public in mid–June, two weeks before the end of the financial year.

Unfortunately, this sound process did not operate in the intended way during the 1970s.[14] The most important points of departure from the ideal were as follows:

1. The forward budget process was not a period of program and policy review, but simply the aggregating of divisional requests, and subsequent arbitrary cutting by the senior finance officer and/or Budget Supply Officer. Part of the problem was the insistence of the Treasury on excessive detail in the submissions, causing the Ministry to spend effort arranging the details instead of reviewing the broad issues.

2. Few field officers bothered to submit budget estimates, and the field estimates that were received at headquarters were ignored in most cases.[15]

3. Ceilings were not broken down within the Ministry by district or by program division. Consequently, many field and divisional requests were simply "wish lists," asking for three or four times the allocation for the previous fiscal year.

4. Budget submissions from parastatals, whose subsidies and subventions took increasingly large percentages of the ministry's budget during the seventies, were not reviewed by the Ministry. Requests were simply passed on to the Treasury.

5. Most submissions from divisions, despite their "wish list" character, simply were passed on to the Treasury also. Only the most obviously padded items were cut before the Treasury discussions.

The budget formulation for MOA in 1980–81 illustrates the consequences of these weaknesses. The Ministry's request to the Treasury was over twice the ceiling, and over twice the spending rate for the 1979–80 fiscal year. The Treasury's Budget Supply Officer was forced to challenge all proposed increases, hear a few sentences of defense, and make an immediate decision on whether or not to allow the increase. Since the budget proposal had been formulated without policy priorities, it could not be well defended on policy grounds. He thus cut deeply, doing so with even less field and expenditure information than was available within the Ministry. The second, internal Treasury review resulted in further, often arbitrary, cuts to accommodate revenue estimates.

In the end, MOA received less than had been projected for it in its Forward Budget. Even worse than the reduced total, however, was the distribution of funds. Important projects and programs were underfunded, while some poorly conceived activities were included. In at least one case, one of two impaired items was cut and the other allowed, when the expenditure only made sense if they were done together. Also, reductions were made by item, not by program. Expenditures such as travel were seen as "controllable" and hence were favorite targets. The obvious result was to hamper much of the Ministry's basic work, since extension and veterinary services depend on travel in order to see farmers and to review field conditions.

Although the budget process is a highly competitive game everywhere in the world, there is a difference between gamesmanship and a failure to play at all.[16] Why did MOA officials play the Kenyan budgetary game in such a cavalier manner? The refusal to establish priorities among programs and projects was due in part to the tremendous expansion in resources during the mid–1970s. Coffee revenues were high, and donors were anxious to put more money into smallholder agricultural development. The correct presumption was that any development project that was well conceived or had donor support would be funded. With regard to recurrent programs, the valid assumption was made that if allocated funds were exhausted well before the end of the year, the Treasury would provide additional money. Far from being disciplined for being financially uncontrolled, officers were rewarded, in effect, with more funds. Thus, officers saw no need to establish the relative merits of projects or to be especially careful in projecting their requirements. When the fiscal position changed at the end of the decade, the new, bad habits persisted.

Preparing Allocations for the Field

Both MOA and the Ministry of Livestock Development (MLD) experienced problems with the issuance of expenditure authority to spending stations. Once the budget was made public, the chiefs of program divisions divided the money in their portion of the budget among the provinces and research stations. Each Provincial Director (PD) subsequently allocated this money to his district officers. In most cases, this second step was not controlled by headquarters.

There were several problems with this method of issuing Authorities to Incur Expenditure (AIEs):[17]

1. Most AIEs were issued late. In MLD in 1981–82, some AIEs left headquarters as late as September. In both ministries, there was another two or four week delay between the PDs receipt of the AIE and his allocation to the districts.

2. Division chiefs sometimes allocated funds in an inefficient manner. There are four possible explanations (which are not mutually exclusive): the chiefs lacked the expenditure and implementation information from previous years which would have helped them to

make proper decisions; they did not consider allocations important since allocations could be changed during the course of the year; they preferred to use a simple allocative device such as giving the same amount to each province because of laziness or political considerations; or they were corrupt.

3. Some provincial directors misallocated funds. Most often, this took the form of the PD misdirecting money for a project in the development budget by giving money to a district which was not supposed to participate in the project. In other instances, an inordinate proportion of the funds were retained by provincial headquarters. As the PD was not required to report his distribution of funds to headquarters, such problems often went undetected.

Expenditure Monitoring

The tracking of expenditures is important both for managing resources during the fiscal year and as an input into the budgeting process for the next fiscal year. The government's accounting system maintains good records of bills that have been paid and of the present cash flow. Most purchases by officers in the field do not involve an immediate exchange of funds, however, but are made by providing the vendor with a voucher, which he subsequently presents to the government for payment. Thus, large amounts are committed by spending officers long before any payments are made. To keep track of these amounts the government requires those with spending authority to make monthly reports on amounts committed and spent. These reports are compiled at headquarters and provided to senior administrators for their review and action.

Unfortunately, the traditional compiled reports were unfocused, inflexible, untimely, and inaccurate.

1. The final report typically filled over 100 pages of tables. While a summary report by program was produced, all numbers were given equal emphasis. Only the most dedicated civil servant would search through this long document to find problem expenditures.

2. The reports were compiled in only one way, so as to give the total commitments of all spending stations on each item in the budget. If a report of the total financial position of each spending station, or the total amount spent on one category such as travel expenses was desired, all of the numbers would have to be reprocessed by hand, with all the delays and errors that this would involve.

3. There were two delays in the system: provincial directors had to collect all of their district expenditure data, then add them together and send a provincial report to headquarters; then the headquarters staff had to add all the provincial and research station reports. The

headquarters addition could not begin until all expected reports were received. The final report was rarely available before the information was two months old. If, in an emergency, it was necessary to process whatever information was available at headquarters and not wait for all of the reports, there would be at least a 10–day delay before the report was ready.

4. Errors occurred throughout the process. Some stations failed to submit returns; others submitted incorrect or incomplete returns; the headquarters clerks made errors while adding numbers; a middle level official who reviewed the final report before typing would routinely "correct" numbers which looked suspicious; and the typists would average at least one error per page.

The consequence of these many problems was that the reports were rarely used. Major over– and under–expenditures went largely unnoticed and, hence, without corrective actions. The main restriction on expenditure was simply the limit on the funds in the Ministry's cash account, so that bills would not be paid when presented, and creditors would slow down or stop the delivery of goods in consequence. Absolutely essential services would be hampered as much by this response as would low priority ones.

Using Microcomputers for Financial Management

By late 1981, the cumulative effect of the problems of budgeting and financial management described earlier had become unsupportable in Kenya. In early 1982, the President appointed a special working party of senior civil servants to recommend ways to reduce and control government expenditure.[18] In MOA and MLD, TAP management advisors and the Ministry's senior finance officer had recognized the importance of these problems and had begun a series of reforms to correct them. Microcomputers were useful in dealing with the three problem areas earlier discussed.

Microcomputers and Expenditure Monitoring

The first use of the microcomputer after the late 1981 demonstrations was in setting up in the MOA an expenditure monitoring system. The microcomputer used was a 64K Apple II+ with a CP/M card, two disk drives, and a dot matrix printer. Several other low–priced microcomputers could have performed the same functions. The advantages of the Apple for use in Kenya at this time were: (1) the large amount of software available, particularly with the CP/M card installed; (2) the Apple's relatively high level of resistance to fluctuations in power supply; (3) the presence of several other Apples in the country, together with the private sector technicians capable of repairing them; and (4) the ability of the Apple to accept expansion boards which substantially increase the capacity of electronic spreadsheet software (VisiCalc).

The expenditure monitoring system was set up by creating several VisiCalc templates for the recurrent and development budgets in MOA. The reports from the field were then entered as they were received, whereas previously the clerks could not begin adding numbers until they had received all expected reports. Since the totals and percentages were recalculated as each figure was entered, a report from all available information was available at any time. VisiCalc proved to be easy to understand, with professionals beginning to use it with less than two hours of formal training if an experienced user were available for consultation as questions arose.[19] The lower–level clerical staff who were used to entering the expenditure reports also learned data entry techniques quickly.

Several of the shortcomings of the traditional expenditure reporting system were lessened directly by this application of the microcomputer. First, final reports were available much more quickly. Preparation time was reduced by two to five weeks, mainly by entering each report as soon as it was received, rather than after all had arrived. This time difference proved crucial in getting decision makers to see the reports as current and, therefore, relevant. Once decision makers were aware of the reports, they were able to request and receive summary information at any time in the compilation process if timeliness was more important than having a complete report from all spending stations.

Secondly, the accuracy of the reports was improved. In the past, significant errors had been made in addition and typing. Human error in calculation was, of course, eliminated by the machine. Most typing errors were avoided by having the data processor enter his own figures. Carelessness by typists, to whom the figures had no intrinsic meaning, had been a major problem in the past, and much professional time had been absorbed in the proofreading and correction of their work.

Thirdly, the compiled reports became somewhat more focused and therefore usable by decision makers. VisiCalc made easy some changes in format design, leading to cleaner and more "transparent" documents. The final reports also highlighted some problem areas of expenditure, but much of this work was done by hand, mainly by TAP management advisors.

Despite these successes, after several months, it was clear that VisiCalc was not the appropriate tool for the Ministry's expenditure monitoring system. There were three problems. First, the flexibility issue had not been resolved, and senior officials were asking for expenditures compiled by spending station and by type of expenditure. With VisiCalc, it was necessary to re-enter most of the data to produce such reports. Second, although the typists had been bypassed, there were still numerous keyboard errors each month. It was desirable to have a program that could check data entry for certain types of errors. Third, VisiCalc's reports were not flexible enough for the numerous ways in which the data were being used. In particular, VisiCalc was unable to highlight problem areas. Finally, the decision was made to use the microcomputer to allocate funds directly to the districts (described as follows), and it was clearly preferable to use one piece of software for both the allocation process and the expenditure monitoring system. A relational database management system was the obvious answer. It is important to note, however, the remarkable fact that a US $112 million organization like MOA was able to use such a simple and limited tool as VisiCalc to improve dramatically the quality of its financial reports.

Since the beginning of the 1982/83 fiscal year, the database management system dBase II has been used for monitoring expenditures. This has solved most of the remaining problems. Timeliness has been improved further, as there is no longer a delay for compilation at the provincial level. A number of validation routines check the data as they are entered to improve accuracy. Also, dBase II allows the programmer to set up data entry screens that look like the printed form. This also decreases keyboarding errors. The main advantage of the database management system, however, is in the ability to focus on any aspect of the system, and produce whatever type of report is required. Each record of the database is identified by a spending station, a program (such as agricultural extension), and a type of expenditure (such as farm inputs). Each of these three keys is used to produce a standard monthly report compiled by station, program, and type of expenditure. In addition, each month, a particular type of expenditure is analyzed for the problems encountered by different programs with this item. Custom reports are available as requested.

The price of flexibility, unfortunately, is increased complexity. dBase II is considerably more difficult to use than VisiCalc for the user of the program. However, dBase has its own programming language which allows someone with programming skills to create formats and prompts on the screen which ask the user specific questions. Such a "menu-driven" system has been installed for expenditure monitoring, making the process of entering data and producing standard reports very simple, more simple than it was for the less satisfactory reports produced by VisiCalc. All this work can be done by a clerk who has no knowledge of dBase. Nevertheless, it is necessary to have an operator who knows dBase programming to produce reports not in the menu, or to make modifications to the menu-driven system. While the operator's level of skill is significantly below that necessary to program a mainframe, such persons require considerable on-the-job training. As the use of microcomputers grows in LDCs, governments will have problems retaining such personnel, as they do with computer programmers. Training and retention issues are discussed below.

The reports produced by this system have been used to address problems of over and underexpenditure during the fiscal year, and used for budgeting and allocation of funds for the next fiscal year, as described in the following section.

Preparing Allocations for the Field

Beginning with the 1982/83 fiscal year, it was decided to issue AIEs directly to the districts, rather than through the provinces. While this would solve the problems of delays and misallocations at the provincial level, it also tripled the amount of work to be done at headquarters, since many more AIEs would be issued. Both ministries therefore decided to use the microcomputer in the process, although the decision to change the allocation procedure was made before the machines had been introduced.[20]

Since over 4000 allocations to spending units would have to be made, it was decided that a mechanical procedure would be used to make preliminary proposals. It was necessary to solicit information from the field on actual expenditures in the previous year, since the expenditure monitoring system in place at that time included only provincial, not district, level data. After receiving this information, one of the TAP management advisors used dBase II to produce preliminary allocations based on the combination of previous allocations and expenditures.

These suggested allocations were then reviewed by finance officers and program division chiefs, who checked for omissions and suboptimal distribution of funds. This review stage was important, as the previous year's allocations had suffered from the problems mentioned earlier. Reactions ranged from those chiefs who made only a few changes to those who altered every allocation for their program. In previous years, few division chiefs had referred to past expenditure records in making allocations.

When the alterations suggested by the division chiefs were received, they were entered into the microcomputer as corrections to the existing program. After a few final checks and adjustments, the allocations were printed and distributed to the field officers. Thus, the working documents became the final, legal AIEs themselves with very little additional work.

The result of the exercise was a more rational set of allocations combined with the timely issuance of AIEs. In MOA, the issuance was earlier than in any previous year, despite the trebling in the number of AIEs to be written. All districts and other spending stations now knew their level of spending authority within two weeks of the start of the financial year.

The allocation process for both Ministries for the 1983/84 fiscal year worked in a similar manner. Draft allocations were circulated before the publication of the final budget, based on the budget levels agreed to at the discussion with the Budget Supply Officer. Only those allocations for budget items which were cut in the Treasury's internal meetings had to be adjusted after the final budget was made public. A final copy of the AIEs was then printed without the delays of waiting for typists and proofreaders. The effect of using the machine was not to displace the work of the clerical staff but to spread it out into periods of reduced time stress, thus averting a bottleneck when the need to issue the final documents was urgent. Furthermore, the microcomputer printer was able to cut all the documents onto mimeograph stencils, thus enabling the Ministry to use a reliable, technically appropriate, and efficient method of duplicating them.[21]

This method has solved the problems of timeliness and misallocations by provincial directors. There is still a problem with some division chiefs incorrectly allocating budget funds. Although expenditure information is now available to them, many choose not to use it. A proposal has been made that future allocations be based on a combination of division chief proposals and district officer proposals. Such a system, however, will not work until district officers are involved in the budgetary process, to which we now turn.[22]

Preparation of the Budget

The five specific problems with budgeting mentioned earlier have proved to be less amenable to microcomputer-based solutions than the problems of expenditure monitoring and allocation of funds.[23] As of mid-1984, the forward budget process is still not effective, field officer budgets tend not to be used at headquarters, and parastatal requests continue to be passed on to the Treasury. Some progress has been made in breaking ceilings down to the Ministry's program division but not to districts (a proposal made by TAP management advisors in late 1982 for breaking down ceilings to districts was considered politically complicated and not acted upon). The microcomputers have played a significant role only in solving the final problem area, the passing on of divisional requests to the Treasury without review.

The first year of use, budgeting for the 1982/83 fiscal year, is the best example. Requests from divisions and parastatals in MOA totaled KL168.4 million versus a ceiling of KL56.1 million (combining the Recurrent and Development Budgets). The decision was made to hold a series of in–house budget reviews to reduce the requests to a level at which the Treasury would take the Ministry's budget submission seriously. To facilitate discussion and to make use of the recently begun expenditure monitoring system, the management advisors used VisiCalc to devise a five–page summary of the two budgets. These tables presented basic financial data for each program, such as 1981/82 Budget authorization, 1981/82 expenditure as of December, and 1982/83 Forward Budget figure, along with columns for program division request and the ministry's submission to the Treasury. For the Development Budget, requests were broken down into those from old projects, new projects, and parastatals. Since each of the three groups was treated somewhat differently, the breakdown helped to shed light on where in–house cuts were taking place.[24]

During the meetings, the financial data were referred to whenever a new line item was discussed. If the present year's expenditure was low, the request generally would be cut. After each meeting, the agreed–upon figures were entered into the microcomputer and new forms distributed at the beginning of the next session. Each form included the total amount agreed upon so far, the percentage of requests that had been granted, and the average percentage cut necessary during the remaining sessions if MOA was to stay within the Treasury ceiling.

These meetings brought the Recurrent Budget to within ten percent of its ceiling, but the Development Budget remained about 80 percent above ceiling. Therefore, it was decided that the senior officers in the Ministry should meet again and decide upon further cuts based upon MOA's priorities. To some extent, this discussion was facilitated by the summary format. Clearly, its simplicity and brevity helped to gain the attention of the officers attending and led to a more focused analysis of particular problem items. Some policy issues were raised in this meeting, and a further KL–6 million was cut from the submission.

Similar summary formats have been used for the 1983/84 and 1984/85 budget discussions, using dBase II. These have been helpful in getting MOA's submissions to Treasury close enough to the ceiling to be considered seriously.

As with the allocation of funds document, another aid to the budgeting process for 1983/84 and 1984/85 has been the ability of the microcomputer to print drafts and then cut stencils directly from the corrected draft. While this may not seem like an important innovation, the MLD experience with the 1982/83 submission proves its worth. The in-house cuts in MLD were completed on the 24th of March, and the Treasury deadline for submission was the 25th of March. MLD was unable to give the Treasury a document until the 26th of April, largely because of typing delays and errors. The Senior Assistant Secretary for Finance of MLD spent almost all his time during these four weeks in the frustrating task of getting the document produced, not in analyzing the proposals and preparing his case for the Treasury. Despite all his work, the final document still contained at least three errors of addition. The use of dBase II in the last two years has solved this problem.

At present, the Ministry is beginning to institute a projects management system. One goal of the system is to convert the forward budget into its proper role as a means of program and priority review.[25] The microcomputers are expected to play a role in this new system.

It also should be noted that the Ministry's innovations have influenced the activities of the Rural Planning Division in the Ministry of Finance and Planning. In late 1982, the Kenyan government declared its commitment to District Focus, a decentralization program intended among other things to improve district level planning and disaggregation of ministerial budgets to the district level.[26] In addressing such tasks, the Rural Development Division's officers and expatriate advisors borrowed greatly from the Ministry of Agriculture's microcomputer based financial administration innovations, thereby spreading them more widely in the government and further institutionalizing them in the Ministry.

Microcomputers and Better Financial Management

The experience described in the preceding section clearly shows financial management improving subsequent to the introduction of microcomputers. How important were the microcomputers to the changes that took place?[27]

The microcomputer is a tool. Like any tool, it is useless if not utilized, and may create a poor product if used without skill. Similarly, a very good craftsman can make a fine product even with inferior tools. The importance of good tools is that they make it easier to do a job and therefore increase the likelihood that it will be done. They also sometimes make it possible for a less-skilled workman to do a better job.

It would be misleading to overstate the impact which the microcomputers *per se* have had on financial management in the two Ministries. Many of the changes that have occurred have been primarily the result of other factors. In fact, the presence of the microcomputers themselves is a result of many of the forces that have been conducive to reform.

First, the original microcomputer was introduced partly because MOA and MLD themselves wished to see improvements. If the atmosphere had been unreceptive to reform, the machine might not have been introduced, or might well have remained unused.

Secondly, a junior economist knowledgeable about microcomputers was hired by the TAP to teach Kenyans and expatriates how to use it. Furthermore, he joined two experienced and trusted management advisors. These three technical assistance staff were under pressure to help improve the financial management of the two ministries, and they used the machine to help them to do so. If the microcomputer had not been available, they almost certainly would have carried out some of the same reforms anyway. Similarly, if they had had less skilled or supportive senior officers to work with, they would not have been able to accomplish much, with or without the machine.

Nevertheless, it is clear that the microcomputer did contribute to the reforms that occurred. Its speed and storage capacity made it easier to obtain certain types of critical information. Its accuracy and formatting capabilities made the data more reliable and easier to understand. Thus, the "cost" of information in terms of time and inconvenience was lowered, and senior officers decided to "buy" more of it. Although labor in general is abundant in developing countries, skilled managers are not. Capital investments that will make this scarce resource more productive are likely to be "appropriate technologies."

Furthermore, the presence of the microcomputers visibly changed the technology of financial decision making. Officers knew that it was possible to do things that had not been done before. While some of the reforms that have taken place might have been possible without the machines, it would have been harder to convince officials that they were desirable, possible, and worth the effort. Similarly, the microcomputers legitimized a role in financial decision making for those who could use it, therefore making it easier for certain reform-minded individuals to participate actively.

Once again, the microcomputers did not cause—or even permit—the reforms that have taken place. They are tools that have made the process of improvement and change easier in a setting that was already conducive to it. Given the difficulties of any organizational reform, however, a tool that makes change easier is a precious commodity indeed.

Observations on the Introduction of Microcomputers for Financial Management in Development Ministries

Lessons learned from this experience in Kenya can be divided into three areas: (1) lessons for project identification and preparation; (2) those for the critical early stages of a project; and (3) those for institutionalizing use of the machines.

Project Identification and Preparation

To say that **microcomputers should only be introduced to solve a substantive problem** may sound elementary, but there are numerous instances of machines descending on an organization without any plan for their use or reason for their purchase. The advertising hype which accompanies the introduction of each new machine is partly responsible for this problem. **The substantive problem must be perceived as such by the local officials**, not only by expatriate advisors and consultants. While a research project that provides microcomputers for expatriate researchers may succeed quite well without the support of such officials, improvements in the area of budget and finance require strong local support. Any links that can be made with successful ongoing projects in this area will be helpful. **Implementers should remain flexible about what problems will be tackled** in order to be able to respond to support from local officials and advisors. This may require delaying some seemingly efficient uses of the machines. However, **once microcomputers have proved their worth in one area, it is likely that senior officials will realize the value of using them in other areas.**

The history of the Kenya project underscores the importance of these points. The original demonstrations of the machines were directed towards economists in the planning division of the Ministry. There was little or no enthusiasm on the part of the leadership of this division. On the other hand, the Ministry's senior finance officer was a capable, reform-minded individual who was excited about the use of the machines, and had a particular problem—the expenditure monitoring reports—which could be solved by using the microcomputers. In addition, the TAP project had been providing advisors in the management area for several years. Through their understanding of the administrative setting and their relationship with the officials involved, they were able to steer the project toward areas of work which were likely to succeed, and to build support among their local counterparts.

The emphasis of the project, therefore, shifted from planning to budget and finance. Eighteen months later, after a number of successes in the finance area, officers in the planning division were requesting that machines be introduced there also. The Ministry of Finance is now introducing microcomputers, partly as a result of the successful introduction of the machines for budget and finance work in MOA. Had the original implementers insisted that their work continue in the planning division, the project most likely would have failed.

A project should remain flexible not only in the area of work but also in the number of machines to be introduced. More than in most projects, it is difficult to predict ahead of time how senior officials will react to the introduction of microcomputers and how quickly it will be possible to begin new systems. Introducing many machines before officials are prepared to accept them can lead to a strong negative reaction and failure of the program. Introducing machines too slowly delays institutionalization of the microcomputers, since constraints on machine time limit the amount of training that can take place and require that the expatriate advisor do most of the work. The latter happened in Kenya. For a variety of reasons, the project had been going on for six months before the second machine arrived, and there was another year's delay before a third machine was purchased. Because the amount of substantive work was more than could be handled on the available machines, the advisor was forced to limit any expansion in areas of work for the computer, suspend all training for months at a time, and did much operational work himself since he was the fastest operator. Turning the system over to Kenyans was delayed significantly by the lack of machine time.

The number of machines to be introduced should depend in part on the operating environment and the quality of support services available. While microcomputers are very hardy compared with earlier computers and even with present-day mainframe computers, problems occur, most often in the disk drives and the printer (the only mechanical devices). Excessive dust, smoke, heat, and humidity are the main culprits. It is necessary, therefore, to identify someone who can handle and deal with equipment failures. In fact, a recent consultant's report to the World Bank recommended that the availability of support facilities be one of the most important criteria in the selection of a microcomputer.[28] If there is no one available locally, the technical assistance program must establish links to external repair services and should consider purchasing a complete back-up system. Even where local repair services are available, a back-up system is not expensive, ensures that there will be no delays because of machine failures, and can help the project remain flexible in dealing with increases in demand.

In most cases, it is necessary to hire someone for an extended period of time whose primary job is to introduce the microcomputers. The introduction of any new technology requires the presence of individuals who understand the technology, can teach others to use it, and can identify new applications of it in the present context. This kind of support is commonly provided by vendors and software houses in the industrialized countries, but is generally not available in developing countries. A project which is introducing microcomputers to professional scientists or economists for standard types of data analysis may need to employ such a "teacher" for only a few days or a week. For budget and finance work, however, the officials involved probably will have no previous experience with computers, and will have only vague ideas about how such machines can be used. Successful institutionalization will require an extended period of interchange of ideas between the "teacher" and officials, of writing and revising programs, and of training staff. In Kenya, this period has been three years. A project which provides microcomputers for budget and finance work but no support other than a standard accounting package or a turnkey system set up by a two week consultant is likely to fail.

There is a danger of the specialist being too enthusiastic. The individual must be able to teach, program in the appropriate language, be sensitive to the poor data situation, and decide on efficient uses for the machine. In addition, he or she must be sensitive to nuances from ministry officials whose support is important to the project's success. This sensitivity must include the ability both to raise and lower expectations when appropriate.

While no academic training will prepare the specialist for all of these roles, an economist or MBA with considerable microcomputer experience is probably the best candidate for introducing these machines for budget, finance, and planning. A computer scientist would be overtrained in programming and undertrained in the tasks to which the machine will be put. On the other hand, an economist or business management specialist will be of use in the machine's substantive tasks. He or she will be quicker to identify the ways in which the microcomputer could assist with the problems at hand and will be fully utilized by working on the problems themselves when the locally trained operators are able to handle most of the computer work.

Early Stages of the Project

After the initial area of work has begun successfully, the implementers should consider other efficient areas of use for expansion.

The choice will be constrained by machine time, implementers' time, and the interest of government officials. The third is by far the most important, as the first two constraints can be relaxed by bringing in more machines or hiring more consultants. After the early successes, **some officials will believe that microcomputers are the panacea for all of their problems.** This reaction forces the implementing agents to make choices, balancing efficient uses of the microcomputer with the need to continue satisfying the needs of those officials who are enthusiastic about the machine's presence.

It is particularly important to think about the ways microcomputers are used in developing countries. While the Kenya project is so small that no conclusions can be drawn, there is a potential for labor displacement. **Efficient use of microcomputers in developing countries will vary considerably from industrialized countries,** where clerical labor is much more expensive. Guidelines for efficient use come from the comparative advantage typists and clerks have over the machines. Microcomputers have an absolute advantage in speed and accuracy of computation, ability quickly to revise formats, and in the rapid retrieval and analysis of large amounts of data. A secretary with a typewriter has a comparative advantage (although perhaps not an absolute one) over the microcomputer for typing letters. A clerk with a calculator similarly has a comparative advantage when using data that will be entered and used only once. These are fairly clear examples. Often there are hard choices to make, however.

Experience elsewhere suggests that word processing activities can easily force out other uses of the microcomputer. The machine offers tremendous advantages in the production of committee reports that are going to go through several drafts. However, if this convenience is purchased at the price of data analysis for financial decision making, it may be undesirable. It most certainly will be a misallocation of resources if analysis is displaced by the production of straightforward letters and manuscripts.

In the context of a developing country, a good rule-of-thumb would be that any task which is already being done fast enough to serve its most important purposes should not be done by microcomputers. The machines should be reserved for those uses that have been left undone with the existing technology, or have been done so slowly or so inaccurately that they fail to meet the purposes for which they are needed. These are the tasks for which comparative advantage is relatively certain and which increase the efficiency of the scarce labor of senior managers. In Kenya, there are such uses for which this new technology can thereby be defined as "appropriate."

In addition to wanting to introduce the machines for inefficient uses, some **overenthusiastic officials are blind to possible errors** produced by the microcomputers. Several times, officials have said something like, "the numbers came from the computer so they must be correct." Of course, errors can and do arise in the process of operating the system. While the machines occasionally have a disk read error or a chip malfunction, the vast majority of errors are program bugs, data entry errors, or source data errors.

When the VisiCalc-based expenditure monitoring system was just beginning, a senior officer wanted to test the accuracy of the new system and discover what kinds of errors were likely to arise by compiling the same report both by hand and on the microcomputer. So a clerk with six years' experience compiling expenditure returns computed the December 1981 returns by hand after the same work had been done on the machine. He discovered 53 budget items where the microcomputer report was incorrect. Of these, 49 can be categorized as source data errors, which the experienced clerk recognized and the inexperienced microcomputer operator did not. The other four differences resulted from data entry errors. In this particular case, there were no program bugs or computer errors.

Thus, reports printed by the computer can be in error. Steps have been taken to minimize errors in the reporting system, and the present dBase II system would catch all of those 53 errors. Nevertheless, accurate reports at headquarters require that field accountants file accurate returns. It is hoped that feedback to the field on the importance of the returns will impress upon the accountants the necessity of providing good data. (In this project, handwritten forms are more likely to be accurate than typed ones, as they will have been filled in by the reporting officer and will not have typist errors. It is difficult, however, to break the habit of what a "good" report looks like). Meanwhile, it is useful to encourage senior officers to question statistics which surprise them and not simply to accept machine-processed reports on faith.

Institutionalization Issues

This project has been successful at beginning to institutionalize the use of microcomputers for financial management. Institutionalization will be complete when all of the following have taken place: (1) Kenyans themselves *demand* the output from the machines; (2) all *operations* are done by Kenyans; (3) all *programming* is done by Kenyans, possibly by local consultants; (4) good microcomputer operators are *retained* by the civil service; and (5) maintenance and supplies are provided for in the budget.

Demand comes primarily from Kenyan officers at present and has continued despite the transfer of several finance officers. It would be difficult, albeit possible, for a new senior finance officer to disregard the computerized expenditure monitoring system and budgetary system, and revert to the old way of doing things. First, the unit which previously compiled the expenditure reports no longer exists. Secondly, and more importantly, the old system could not possibly handle the new one's responsibility for allocating funds. Thus, the **technology has been embedded in the system**, and any officer transferred into the Ministry will have to use it at least for allocation of funds, regardless of his personal views on computerization.

Transferring operations to local personnel has also been successful. The difficulty of making this transfer depends on the software used. All of the applications described in this chapter have involved the VisiCalc or dBase II package programs. Other programs used by this project, primarily in the Planning Division, include ones for graphing (VisiPlot), statistical analysis (MicroStat and MicroTSP), and linear programming (Optimizer). One of the most important advantages of microcomputers is the availability of such mass-marketed, "user-friendly" programs at moderate prices, thus escaping the need for accomplished professional programmers or extensive staff training. When using this type of software, **secondary school graduates can be trained to operate microcomputers**. The qualifications of the graduate and the amount of training required will depend primarily on whether or not local personnel will need to understand programming in dBase.

Carl H. Gotsch has sharpened thinking on training issues by identifying three skill levels.[29] The first level is for clerks who enter data into the system. Compared to VisiCalc, dBase II actually lowers the training needs for this level, with clerks being trained in thirty minutes. Level two personnel are able to manipulate VisiCalc and dBase II interactively. General secondary-level education and above-average mathematics skills seem to be adequate qualifications for training at this level. These skills can be acquired by tutorials, supplemented by personal instruction or a custom-written manual on the types of requests normally made on the system. Numerous books, diskettes, and video cassettes are available for help in teaching VisiCalc and interactive dBase. It is important to note, however, that these aids must be in addition to, and not a substitute for, hands-on experience. Any project which intends to teach local staff must provide plenty of computer time for learning.

Level three personnel understand dBase II programming and can modify existing dBase systems. Training at this level is much more difficult. The transition from using dBase interactively to using it as a programming language is the major hurdle. A recently written tutorial manual should help students to make this transition. At this point, the main conclusion that can be drawn concerns minimum qualifications for success. This project has attempted to train seven Kenyans to level three. Five of those have become discouraged and quit. Only one has been trained to level three so far, while the other remaining trainee is approaching this level. The two who have continued with the training have both had higher secondary school training in mathematics (Advanced Level examination) and no university training. The five who did not succeed all have a lower level of formal schooling (Ordinary Level examination). These qualifications, therefore, appear to be a minimum for level three personnel.

The number of failures, however, is much less important than the quality of the successes. The most advanced of the two trainees has been working with the project full time for 15 months and now does 90 percent of the training of lower-level personnel, and (in conjunction with level two personnel) virtually all day-to-day operations. These include monitoring the system of dBase programs and writing small programs to meet special requests.

The mathematics training at higher secondary level is not in itself a prerequisite to dBase programming. There is no particular item in the higher secondary level syllabus that teaches students to be more adept at using a microcomputer. The attained educational level is only a signal that this individual has sufficient mathematics abilities to be able to consider a problem and to figure out how the microcomputer and available software can be used to solve it.

Some donors have encouraged the project to train senior Ministry officials to use the machines themselves. This seems reasonable when looking at the American experience where many high-level managers have microcomputers sitting on their desks. On reflection, however, the experience does not translate well to financial managers in Kenya. First of all, virtually no managers in Kenya know how to type, and are quite embarrassed to sit at a keyboard and pick out letters. Second, there is very little information that they can get by interacting with the machine that a lower-level civil servant cannot retrieve for them just as quickly. Thus, it seems that **senior finance and budget officers are unlikely to use microcomputers themselves.**

The situation is quite different for economists in the planning division, who clearly gain by using a statistical package themselves. Our efforts at giving "hands on" experience to more senior officials therefore are focused on economists.

Thus, this project has succeeded in meeting institutionalization criteria (1) and (2). The remaining three criteria have yet to be met. To date, most programming has been done by a TAP management advisor. All programs, however, have been written with the Kenyan operators in mind, and an attempt has been made to keep the programs simple and clear. Efficiency of computer time has at times been sacrificed for clarity of presentation. The goal of the project is to identify a local consultant who will be available to assist the normal operators when they encounter a situation beyond their capabilities.

Retention of staff is a serious problem. Microcomputer use is expanding in the developing countries, and operators will be needed in the private sector. At present, there is no government mechanism for promoting skilled microcomputer operators in Kenya. While most civil service systems have schemes of service for programmers, few fully appreciate the skills of someone who knows dBase II and VisiCalc, but not BASIC or FORTRAN. Discussions have been held with the Director of Computing Services for the government, bringing the problem to his attention. He has the authority to promote persons with microcomputer operating skills. Only time will tell if he will do so. If he does not, there is a strong possibility the rapidly growing private microcomputer sector will be able to bid away those trained to level three. While this offers unprecedented opportunities for the advancement of women formerly frozen at the typist–clerical level, it makes it difficult for projects such as the one described here to fully institutionalize their achievements.

Government payment for maintenance and supplies is the final area of institutionalization. As long as a donor–aided project is providing funds, there is no guarantee that the government will take over such expenses at the end of the project. There are plans to introduce a line item in the budget for these expenses at least one year before the end of the project to address this problem.

Conclusion

The uses of the microcomputers which have been described here broke new ground in Kenya. They also were critical to initiating and executing a process of improvement in a governmental system of financial management which had deteriorated significantly in recent years. There is every reason to believe that the careful application of microcomputer technology may help to achieve similar reforms in other countries with management problems. It is hoped that the lessons learned from this project will be of assistance to donors and government officials who are considering applying microcomputers to these tasks.

Notes

1. International Bank for Reconstruction and Development, *Accelerated Development in Sub-Saharan Africa: An Agenda for Action* (Washington, D.C.: The World Bank, 1981).

2. The World Bank's *World Development Report 1983* (New York: Oxford University Press, 1983) takes as its central theme the need to promote more efficient management practices in operating ministries of low-income countries. See also: Coralie Bryant and Louise G. White, *Managing Development in the Third World* (Boulder, Colo.: Westview Press, 1981); Food and Agriculture Organization, *A Programme on Improving the Organization and Administration of Agricultural Development* (Rome: FAO, Report W/N9070, September 1980); Guy Hunter, *The Implementation of Agricultural Development Policies: Organization, Management and Institutions* (London: Overseas Development Institute, 1974). For references with some analysis of budget and finance issues, see: Robert Chambers, *Managing Rural Development: Ideas and Experience from East Africa* (Uppsala, Sweden: The Scandinavian Institute of African Studies, 1974); A. T. Mosher, *Serving Agriculture as an Administrator* (New York: Agricultural Development Council, 1975).

3. David K. Leonard, *Reaching the Peasant Farmer: Organization Theory and Practice in Kenya* (Chicago: University of Chicago Press, 1977); Christopher Trapman, *Change in Administrative Structures: A Case Study of Kenyan Agricultural Development* (London: Overseas Development Institute, 1974).

4. Dwight Brothers, "Management of National Agricultural Development Programs," in *Education and Training for the Public Sector Management in Developing Countries*, L. D. Stifel, J. S. Coleman, and J. Black, (eds.) (New York: The Rockefeller Foundation, 1977); John Howell, "Ministries of Agriculture and Administration of Agricultural Development" (London: Overseas Development Institute, Agricultural Administration Network Paper No. 4, 1980) and his "Training Managers for Agricultural Development Projects" (Undated paper, Overseas Development Institute, London).

5. Carl H. Gotsch, "Report to the Permanent Secretary on the Potential Benefits of Selected Computerization in the Ministry of Agriculture," (Report by Gotsch Associates, Inc., Palo Alto, Calif., January 1982).

6. Thomas C. Pinckney, John M. Cohen, and David K. Leonard, *Microcomputers and Financial Management in Development Ministries: Experience from Kenya* (Cambridge, Mass.: Harvard Institute for International Development, Development Discussion Paper No. 137, August 1982); David K. Leonard, John M. Cohen, and Thomas C. Pinckney, "Budgeting and Financial Management in Kenya's Agricultural Ministries," *Agricultural Administration*, XIV, 2 (1983), pp. 105–20; and their "Microcomputers and Financial Management in Development Ministries: Experience from Kenya," *Agricultural Administration*, XIV, 3 (1983), pp. 151–67.

7. World Bank, *World Development Report 1983*, pp. 72–3; Marcus Ingle, Noel Berge and Marcia Hamilton, *Microcomputers in Development: A Manager's Guide* (West Hartford, Conn.: Kumarian Press, 1983), pp. 120–3.

8. This paper deals with improvements in financial management which were aided by the microcomputers. It does not describe all microcomputer applications, nor does it describe all improvements in financial management. For the latter, see the series of papers written for the Task Force on Budget and Financial Management Processes, available from the Management Systems Unit, Ministry of Agriculture and Livestock Development, Nairobi.

9. Phase II of the TAP is funded by UNDP, CIDA, SIDA, USAID, and IBRD. It funds two management advisers. The IBRD's Agricultural Technical Assistance Project (ATAP) is an add on to TAP II. It funds microcomputer equipment, an adviser on microcomputers, a finance and budget adviser, and task forces on budget and finance under project No. Cr. 1277–KE.

10. Paul S. Haddow, "Program Implications of the Current Budget Allocation to the Ministry of Agriculture," (Memorandum to the Development Planning Division, Nairobi, Ministry of Agriculture, August/September 1980).

11. In Kenya, each ministry has two budgets: Recurrent, for locally funded expenditures of a recurring nature; and Development, for capital expenditures and foreign–aided projects. For an analysis of the patterns of underexpenditure in the Development Budget, see: Joseph J. McGovern, "Underspending of Development Estimates in the Ministry of Agriculture: Seven Case Studies" (Memorandum, Technical Assistance Pool, Ministry of Agriculture, Nairobi, 1979).

12. For a more detailed description, see Pinckney, Cohen, and Leonard, *Microcomputers*, pp. 12–15; Leonard, Cohen, and Pinckney, "Microcomputers and Financial Management," p. 107.

13. For a more detailed description, see: W. B. Cutter, "Mission to review financial management issues in the Ministry of Agriculture" (Memorandum to EAPCA, The World Bank, Washington, D.C., 20 April 1981), and Beat Alexander Jenny, "The Administration of Planning and Project Preparation in Kenya," in *Development Planning in Kenya*, T. Pinfold and G. Norcliffe, (eds.) (Downsville, Ont.: York University Geographical Monograph No. 9, 1980), pp. 16–20. A flow chart of the budgetary process is included on page 18 of the latter, and reproduced on page 11 of Pinckney, Cohen, and Leonard, "Microcomputers and Financial Management."

14. A more detailed description of the reality of agricultural budgeting is given in Pinckney, Cohen, and Leonard, *Microcomputers*, pp. 15–18, and Leonard, Cohen, and Pinckney, "Microcomputers and Financial Management," pp. 109–112.

15. Michael Chege and Joseph Kimura, "Budgetary Estimate Preparation at District and Provincial Level: A Case Study of the Ministries of Agriculture and Livestock Development in Meru District, Eastern Province" (Paper prepared for Ministry of Agriculture, Management Systems Unit Task Force on Budget and Financial Management Processes, Paper No. 4, April 1982).

16. Aaron Wildavsky, *The Politics of the Budgetary Process*, 3rd edition (Boston: Little, Brown and Co., 1979).

17. A "spending station" is any station in the Ministry where there is an officer with authority to sign payment vouchers. These include district agricultural offices, research stations, and farmer's training centers, among others. An AIE is the authority under which the vouchers are written. Each AIE is for a particular type of expenditure, and limits the amount of money the officer can spend on that item. For a more detailed description of the problems of issuing AIEs, see: Thomas C. Pinckney, G. K. Githae, and Z. E. N. Achira, "The Allocation of Funds for 1982/83 in Kenya's Ministries of Agriculture and Livestock Development" (Paper Prepared for Ministry of Agriculture, Management Systems Unit Task Force on Budget and Financial Management Processes, Paper No. 7, October 1982).

18. Many of the issues discussed here are touched on in the committee's report and recommendations made for corrective action. Philip Ndegwa, (Chairman), *Report of the Working Party on Government Expenditure*, (Nairobi, Government Printer, July 1982).

19. We have presented elsewhere a more complete description of the VisiCalc program and the applications mentioned here. See Pinckney, Cohen, and Leonard, "Microcomputers and Financial Management," pp. 28–35.

20. Both MOA and MLD used similar procedures. Only the MLD experience is described, even though MOA made the change first, as the process was further refined and somewhat simpler there. Note that MLD was split from MOA in 1979, and then the two were merged again into the Ministry of Agriculture and Livestock Development in September 1983. The microcomputer innovations described here began when the ministries were divided. But TAP management advisers in both ministries did everything possible to keep the use of the new technology parallel in each ministry, a strategy which was rewarded when the ministries merged.

21. At about the same time as the switch was made from VisiCalc to dBase II, the Ministry received a second system which included an impact dot–matrix printer rather than the thermal printer which came with the first machine. Since the new printer could cut stencils, the expenditure reports were no longer dependent on the Ministry's photocopy machine, which was often broken or out of paper or ink, thus reducing an important technical constraint and matching the microcomputer technology to the mimeograph based systems of the Kenyan government.

22. Pinckney, Githae, and Achira, *op cit.*, "Allocation of Funds."

23. For a general discussion of the budgeting situation in the recently merged Ministry of Agriculture and Livestock Development, see: D. K. Ole Nasieku, Ole Nils Isaksson, and S. Ramakrishnan, "Review of Present Budgetary Processes and Measures for Improvement" (Paper prepared for Ministry of Agriculture and Livestock Development, Management Systems Unit Task Force on Budget and Finance Management Processes, Paper No. 18, January 1984).

24. Copies of the formats used are printed as annexes to Pinckney, Cohen, and Leonard, *Microcomputers*.

25. Lester Gordon, "A Proposed Projects Management System" (Paper prepared for Ministry of Agriculture and Livestock Development, Management Systems Unit Task Force on Budget and Financial Management Processes, Paper No. 19, January 1984).

26. Documents on these policies and innovations are found in: Kenya Rural Planning Project, "Rural Planning Project: Progress Report January–August 1983" (Nairobi: Ministry of Economic Planning and Development, 1983).

27. Parts of the following section draw from a consultant's evaluation. Carl Gotsch, "An Evaluation of the Present Contribution and Future Potential of Microcomputers in the Kenyan Ministries of Agriculture and Livestock Development" (Report prepared for Ministry of Agriculture, Harvard Institute for International Development, and

World Bank, Stanford University, Food Research Institute, Palo Alto, Calif., August 1983).

28. Jeffrey Stone, Hal King, and Mary Owens, *Microcomputer Selection Study* (Washington, D.C.: The World Bank, Computing Activities Department, July 1981).

29. Gotsch, "Evaluation." These levels were expanded subsequently to four; (see Chapter 3.)

6

Microcomputers for Improved Government Budgeting: An African Experience

Clay G. Wescott

Introduction

Kenya's 1985/86 Budget[1] was produced by the Ministry of Finance and Planning (MFP)[2] using microcomputers for the first time. The project was extremely ambitious, and is the only case we know of worldwide where a national government budget has been prepared in this manner.

The Ministry acquired its first microcomputers in 1983. At that time, the Government was facing rapidly expanding data processing needs coupled with restrictions on hiring additional staff to meet these needs. There were two dimensions to these expanded needs: the volume of data needed by decision makers was increasing, while the time available for processing these data and preparing reports was being reduced.

The volume of data was increasing as the Government took on broader responsibilities to serve an expanding population. For example, the District Focus Policy[3] required that budgetary and other financial data on Government operations be disaggregated to the District level, multiplying by many times the volume of data that had to be

Readers wishing more detailed information are referred to Development Discussion Paper No. 227 of the Harvard Institute for International Development, Cambridge, Massachusetts, April 1986, "Microcomputers for Improved Budgeting by the Kenya Government."

handled. In another example, periodic reporting and compliance requirements of various donor conditionality agreements required processing to a high degree of precision data on nearly 600 donor–assisted projects, in a limited amount of time. Traditional, manual methods of data processing were not adequate to meet these and other new requirements.

Computing technology in use throughout Government was also unable to keep up with the expanded data processing requirements. The MFP's Government Computer Services Centre (GSCS) was and is responsible for all official computing activities. Because of the volume of work, and because of staff vacancies caused by excellent job prospects outside Government service for computer programmers, the GCSC's data processing capabilities are lagging far behind the Government's needs.[4]

A new approach was suggested by the Ministry of Agriculture and Livestock Development (MALD). In response to prodding from the MFP to improve its financial reporting systems, the MALD introduced microcomputers in their Finance Division in late 1981, with the support of three expatriate advisors. Microcomputer–assisted systems were developed for expenditure monitoring, allocation of expenditure authorization, and budget preparation.[5]

The experience at MALD was well documented and seemed promising for MFP. Microcomputers seemed to be robust in this environment, and Kenyans without any background could be trained to use them. Many previous data processing problems were solved, staff morale increased, and scarce Kenyan professionals and technical assistance personnel were becoming more productive.[6]

The same contractor, the Harvard Institute for International Development (HIID), was responsible for the Rural Planning Project, Phase 2 (RPP2)[7] in the MFP. RPP2 staff were concerned that progress in implementing the district focus policy would require significant improvement in the Ministry's data processing systems. In February 1983, the new RPP2 Management Advisor completed a detailed feasibility study examining the potential uses of microcomputers in carrying out the District Focus Policy. As a result of his findings, the Ministry decided to purchase four Osbornes and one Apple microcomputer.[8]

The microcomputers introduced under the Rural Planning Project were used initially in the production of 40 1984–88 District Development Plans.[9] Whereas the previous District Plan production effort had required over two years to move from first drafts to publication, the most recent effort was completed using microcomputers in six months, including time for training secretarial staff. In other parts of the Ministry, recently trained Kenyan staff and locally hired consultants transferred The Weekly Exchequer Return (which monitors the Government's overall financial position each week) to the microcomputers, permitting faster turnaround and more detailed information than was possible using a manual system. In yet another application, in response to a severe drought, models were developed on the microcomputers to estimate food requirements for each district, schedules for unloading ships and transporting food to depots, subsidy requirements for the National Cereals and Produce Board (NCPB), and the national budgetary impact of these subsidies.

A new group, the Management Information Section (MIS), was set up by the Director of Planning to ensure the optimal utilization of the computers, in close coordination with GCSC,[10] and with technical assistance from the Management Advisor.

After a trial period of six months, senior officials were asking for more than the initial units could produce. The MIS had developed two major databases: the Development Project Status Reporting System, with monitoring data collected from Ministries on over 1400 projects, and the *Compendium on Development Cooperation with Kenya*,[11] with data collected from donors on nearly 600 projects. Over 30 major new applications were requested by officers throughout the Ministry. The most challenging requests were from the Budget Department, as will be discussed shortly. However, there was a reluctance to approach donor agencies for additional machines because of the political sensitivity of the issue (senior government officials had made well-publicized speeches that computers and electric typewriters were inappropriate for Kenya), and because GCSC was resisting additional microcomputers not under its direct control. A compromise solution was reached: the Management Advisor would write a proposal to the Kaypro Corporation asking for a donation of ten units under its corporate grants program. The proposal was accepted, and the new units arrived in May 1984. By that time, the political climate had become more favorable.[12]

Preparation of 1985/86 National Budget

Objectives

The objective of computerizing the National Budget was to increase the productivity of the Budget Officers by simplifying the mechanics of budget preparation by performing all calculations automatically, lowering the error rate, reducing the need for retyping, and skipping the laborious typesetting phase. In the past, the typing of the Estimates had been a tedious and time-consuming activity. The process had been error-prone and subject to constant revision, thus requiring several cycles of retyping. The typesetting phase had also been very cumbersome, and required Budget Officers to spend many late nights going over proofs at the office of the Government Printer. The greater disaggregation of budgetary data required under the District Focus Policy intensified these problems. As a result, Budget officers spent a disproportionate amount of time dealing with the mechanics of production, and did not have sufficient time for thoughtful budgetary analysis.

Under the computerized system, headings and initial data are entered only once, then the database can be revised at will simply by typing in the corrections. All totals automatically adjust, and a clean copy of the entire report then can be printed out again. With the mechanics of budget production thus under control, professionals have both the time and the processing power to carry out budgetary analysis, to fine-tune budget structure and projections, to make thoughtful revisions of investment priorities to meet changing revenue situations.[13]

Program of Work

In March 1984, a locally based expatriate consultant was hired by the Management Advisor to write a database management program (in dBase II) for producing sample portions of the 1983/84 Annual Estimates. Such a programme had already been written by a Harvard advisor at the MALD, but the resultant printout had a different format from Government's printed estimates, and was confusing to the MFP Budget Supply Officer concerned. The consultant was successful in matching the format of the printed Estimates, and on this basis, the Financial Secretary ordered that the entire 1985/86 Estimates be prepared using microcomputers.

Over the next 15 months, a team of 12 data entry and programming staff, four Kenyan consultants, two expatriate advisors, and three expatriate consultants worked with the Budget Department to computerize the 1985/86 Estimates. The critical point in the exercise came from mid-April to early June 1985 (see Table 6.1), when the entire team worked on all available microcomputers in the Ministry around the clock to ensure that the exercise was completed on time. Since most of the data were presented for processing during the last two weeks of April, the microcomputer staff initially was overwhelmed. In a number of stormy meetings, the Budget Department complained of the high rate of errors, both from malfunctioning programs and typographical mistakes. They also complained of long turn-around times.

The Management Advisor, acting as a spokesman for the microcomputer team, explained to the budget officers that:

Table 6.1

Timetable for Preparation of 1985/86 Estimates

March–June 1984: Initial dBase II program development

April–May, 1984: 10 Kaypro microcomputers received by Ministry; Kenyan programmer appointed by GCSC to head up exercise; he undergoes training in microcomputer operations.

June–December, 1984: Microcomputer section established; located in room close to Budget Department; Head trains staff of four Ministry copy typists to use Kaypros running dBase II to enter data for 1984/85 Estimates; this exercise used to further develop and debug programs; Head also trains a designated Budget Officer in microcomputer operations so he can manage effectively data-entry staff and provide liaison between budget and computer staff.

July 1984: Financial Secretary and Management Advisor meet with Government Printer to explore possibility of producing the 1985/86 Estimates by offset printing of microcomputer printouts. Printer agrees.

August 1984: Ministry requests new microcomputer equipment and technical assistance from USAID; request is accepted.

January 1985: Budget Officer assigned to provide liaison with microcomputer section is sent away for six month training course (unrelated to this exercise); long delay in appointing a replacement.

January 1985: New USAID contractor, Thunder & Associates, begins work on non-budgetary applications; first two IBM PCs installed in Ministry.

February–March 1985: Microcomputer staff scheduled to enter Revised 1984/85 Estimates; however, data from most ministries delayed.

April–June 1985: Intensive work period as data for nearly all ministries submitted to computer section for data processing; all microcomputer staff, consultants, and equipment deployed on budget work.

2 June 1985: Final printouts from 1985/86 Estimates delivered to Government Press for offset printing.

13 June 1985: Minister presents budget to Parliament. 15 July 1985: 1985/86 District Budgets printed and delivered to District offices.

1. Most data entry staff were copy typists unfamiliar with budget formats; higher-level budget secretarial staff had not been made available to do computer work;
2. No one had been appointed from the Budget Department since the departure of the Budget Officer appointed to manage the microcomputer operation;
3. Hand-written corrections by budget staff on printouts were often illegible, due to complex networks of circles and arrows.
4. For some ministries, the heads and/or subheads in the Estimates were not in numerical order. The Budget Officers concerned insisted that this order be preserved. In order to preserve this order, the computer could not be used for automatic sorting; this meant that the program for preparing the votes for these ministries was much more difficult to use than originally envisioned. It also meant that there were many different programs that had to be used depending on the ministry, leading to confusion among the data entry staff as to which program should be used on a given vote.
5. In the Development Estimates, some ministries insisted on subhead totals. Other ministries insisted on not having subhead totals. Thus, different programs had to be prepared, again leading to confusion by the data entry staff as to which program should be used for which vote.

Given the ambitious nature of the exercise, it is not surprising that jurisdictional disputes arose. For example, one of the longest and most difficult votes was for MALD. This was ironic, since MALD had been the first ministry to computerize its budget. However, since their system was incompatible with the system at MFP, data had to be re-entered. Due to the pressure of work, the Management Advisor asked MALD if they could help out with data entry. This led to compound misunderstandings involving staff at MALD (they had other pressing work for their microcomputer staff, and were worried that MFP might permanently hijack their staff), the Budget Officer concerned (who wanted the head MALD programmer to make modifications in MFP's program which she wasn't capable of making), and the MFP microcomputer team (who needed both data entry assistance and substantive input from MALD to explain the idiosyncratic format of their budget). The resultant acrimony seriously delayed progress on MALD's budget.[14]

The management of the microcomputer operation proved more difficult than anticipated. From June 1984 to mid-April 1985, day-to-day management was carried out by the Head of the Microcomputer Section and the designated Budget Officer (until January 1985 when he was sent for training).[15]

This arrangement worked reasonably smoothly until the pace of work stepped up in mid-April. By that time, the Head of the Microcomputer Section was fully occupied writing new programs, and solving problems in existing ones. Thus, starting 22 April, the management advisor and the expatriate microcomputer consultants took over full-time managerial responsibilities, including:

1. Receiving all new work from the designated budgetary officer and other budget officers, and logging it in; noting any unusual problems; returning work to officers without any corrections being made if estimated totals were over the ceilings (difficult since some of the ceilings kept changing).
2. Assigning work to staff; ensuring that the correct programs for data entry and texting (producing a text file from the revised database for printing) were used for each job; answering technical questions if possible; locating the Head of the Microcomputer Section or the Microcomputer Advisor to answer questions if necessary.
3. Ensuring that computers were not used for texting and printing if there were idle data-entry staff and a backlog of data entry work; these automatated tasks were saved for periods when there were more computers than staff, such as lunch hour and evenings.
4. Ensuring that sufficient staff were on duty so that all machines were in use at all times; searching for missing staff throughout the building; trying to prevent other Ministry officers and advisors from assigning nonbudget work to computer staff;
5. Designing management systems for handling the paper-flow, including blackboard status charts on all votes, and a work-in-process filing system to help keep track of work as it moved through the different stages of production.
6. Proofreading printouts; making decision to release to Budget Department or return to computer staff for further corrections.
7. Arranging for tea to be available for the computer staff and for lunch to be delivered to staff working on weekends and holidays; the Management Advisor donated a teapot. This was all part of an effort to improve the workplace environment in a period of stress.
8. Arranging with the Ministry to provide transport to staff working after normal hours.
9. Negotiating with the Budget Department over changes in format to expedite data processing.
10. When not doing any of the above, assisting with data entry, corrections, texting, printing, equipment repair, or whatever other odd jobs needed to be done.

A Kenyan consultant was hired to clear up the backlog each evening. Starting from 7 pm when data-entry staff started going home, he backed up all disks, and cleared the texting and printing files whenever possible. These jobs kept him working some nights until 5 a.m.[16]

The acrimony between the computer section and the Budget Department reached a climax on Friday, 10 May. Following reports of vocal dissatisfaction by some Budget Officers, the Head of the RPP2 and the management advisor went to see the Permanent Secretary to plead for more cooperation, and for a final decision on ceilings so that the Recurrent Votes could go to press. The Permanent Secretary said that except for the Financial Secretary and the Chief Estimates Officer, the entire Budget Department had moved from total support of the exercise in March to opposition. He said that if the situation did not improve dramatically within a few days, that he would arrange personally for the Development Estimates to be taken from the computer section and prepared in the traditional manner.

Fortunately, there was a dramatic improvement. Within 1/2–hour of the meeting, the ceilings were approved and the computer section authorized by the Chief Estimate's Officer to run final prints of the Recurrent Estimates. The Chief Estimate's Officer also authorized numerical ordering of budget heads, vastly simplifying the production process. The first recurrent votes were sent to the Government Printer on 14 May, and the final page on 18 May. The first bound copy was delivered on 20 May. This led to a major improvement in nearly everyone's spirits. What had seemed a debacle to the Budget Department a week before now seemed a remarkable achievement. The final pages of the Development Estimates went to the Government Press on 4 June, a Ministry record for promptness in the current decade.

Attention then shifted to District Allocations (or budgets). Because of unexpected programming complexity in producing the National Estimates as discussed earlier, the programs had not been prepared for printing out District Allocations in the required format. Furthermore, because of delays in the National Estimates, work on the 1985/86 District Allocations couldn't begin until June.

For these reasons, the microcomputer management team suggested to the Chief Estimates Officer a different approach. He agreed to give it a try. He asked the Management Advisor to prepare templates for each section of the District Allocations using a word processor. The Government Press then produced sufficient copies of these templates using offset printing. Ministries were used to type in their allocations. The completed forms were reviewed by the Treasury, corrected as needed, and sent to the Government Press for a photo–reduction and offset printing.

The previous District Allocations had been typeset, half by a private firm and half by the Government Press. Because of production delays, they had not been released until October and February respectively, too late to be of any use. The 1985/86 Estimates were printed and sent to all Districts by mid–July, in time to be fully utilized by districts in their annual expenditure and budgetary cycle.

Lessons Learned from the 1985/86 Budget Exercise

The production of the 1985/86 Budget has been a major achievement. The result of more than a year of careful planning has been the earliest completion date in many years, and much less tedious work by the Budget Department (although some of the work was shifted to the Microcomputer Section). This is the first time in the developing world that a national budget has been produced entirely on microcomputers. However, the design of the 1985/86 budget process should not be emulated.

Problems

It is not surprising that such an ambitious exercise experienced many problems. Some of the major ones were:

1. There was not enough communication between the computer staff and the budget officers. The reason in part was the assumption that there was no point discussing the exercise with Budget Officers in advance; they would resist at first, but in the end, see that the new system made their work much easier. With hindsight, it is clear that an orientation/question session should have been held at the beginning so that each side understood the other's problems and method of working. The Financial Secretary and the Chief Estimates Officer were both involved at every stage in the design of the exercise, and they remained consistently the effort's staunchest supporters. If some of the other Budget Officers had been involved in the early stages, they might have felt more a part of the exercise. Such communication in the future will be essential if computerized budgets are to become institutionalized.

2. There was also not enough communication within the computer staff. The staff needed a better understanding of what they were doing, why they were doing it, what others expected from them, and their current status toward meeting these expectations. There should have been a more formal system of getting all budget staff together to discuss problems, and to agree on necessary changes in policies to solve them. During the exercise, programs, disks, and various policies were constantly changing. Informal adjustments by word of mouth, adhesive stickers, and illegible scribbles on the blackboard were not an efficient method for communicating these changes.

3. Such communication problems are endemic in mainframe computer operations which physically separate "computer experts" from end users. Since the microcomputers were also physically separate from end users, the exercise had a close resemblance to mainframe computing with all the attendant problems. In addition to the communication problems, this approach was responsible for tying up too many staff members, and for considerable duplication of work: the numerous iterations of printing, writing amendments, keying in amendments, retexting, and reprinting.

4. Although there was a clear strategy from the beginning to guide this exercise and extensive preplanning, management procedures and roles were changing constantly during the intensive, six-week period. It was hard to anticipate this because no one involved had any experience in a microcomputer exercise of this magnitude. Yet the result was confusion and tension among the computer staff and in relations with officers and the Government Press. Again with hindsight, it would have helped to spell out more clearly the roles and responsibilities of all the actors. Management procedures at all points in the process must also be understood by all the actors concerned.

5. The 1985/86 budget process was too technical. Few programs were menu-driven, no documentation was available, and no one person fully understood the whole process. These problems resulted in large part because the job proved to be too big for dBase II. Yet more powerful database programs such as dBase III hadn't even been released when this exercise was conceived in March 1984.

6. There were too many expatriate advisors and consultants involved in the process. The roles played by these advisors could have been taken by Kenyans, were trained officers available. Of the five expatriates involved, four were largely concerned with management and supervisory responsibilities.

Some of these problems are serious ones; a great deal of thought is going into how to solve them during the next budget cycle. Yet despite these and other problems, the nearly unanimous view of all the participants is that the exercise was a success.

Principles in Introducing Microcomputers

From the experience of this budget project—its success and its problems—there may be distilled some basic principles for introducing microcomputers in an African government setting.

1. **Senior Management must recognize the benefits of improving data processing efficiency, and have an open mind toward innovations which achieve this.** Six months prior to the introduction of

the Ministry's first microcomputer, the *Working Party Report* found "...many instances of mismanagement, inefficient organization, failures to follow established procedures, and a lack of diligence in implementation...",[17] and offered detailed suggestions for improvement throughout Government service. Although the report makes no specific recommendation concerning microcomputers, many of the improvements identified would be difficult to achieve without the use of computers.

Even though senior management was already looking for improvements in efficiency, at first they opposed the use of microcomputers on the grounds that: (1) microcomputers were politically sensitive in that important Kenyan politicians and civil servants believed that computers would displace workers, that they were too sophisticated for Kenya's stage of development; and (2) the GCSC itself resisted the use of microcomputers over which it had no direct control.

Despite these objections, the overriding desire of MFP senior management to increase efficiency prevailed. They became persuaded that microcomputers should be introduced. Equally important, senior management sustained their support of this introduction in the face of forceful opposition during the 1985/86 Estimates exercise from the Budget Department due to initial misunderstandings and start–up problems within an extremely limited high–pressure environment.

2. Initial applications must be successful, and considered by senior management to be of highest priority. Microcomputer staff should find out which reports senior management need most urgently, and are having greatest problems producing using manual techniques. These should be the first applications for the microcomputer. Reports should be accurate, on time, and as close as possible to the format of reports produced by the previous, manual system. Senior management will be wary, and probably wish to continue a parallel manual system for some time, which they should be encouraged to do.

In the Ministry, there was a long history of unfulfilled promises from the overloaded GCSC; thus, officials were quite properly distrustful about familiar promises concerning microcomputers. It was crucial that officials received visible early proof that microcomputers were different: that **they were much more reliable**; if one were down, others could take its place; that **they were much easier to operate**, a new report could be designed in the morning and delivered in the afternoon. (Of course, some applications on microcomputers are very difficult to implement, such as data processing for a major national survey. These sorts of applications should be avoided during the initial period.)

3. **A core group of enthusiastic promoters is essential.** The MFP effort was initiated by expatriate advisors who had used microcomputers on previous assignments. During the first few months, the advisors built up a group of Kenyan secretaries, clerical officers, Ministry professionals, and local consultants that used the microcomputers extensively. Some started coming to the office 10 to 15 hours a day, unheard of before this equipment was introduced. Indeed, much of the productivity gains stem from the improved motivation of low-grade staff.[18]

The Rural Planning Project has helped to stimulate this new motivation by selectively paying overtime, but this does not seem to be the prime motivator for top staff members. The root of the new enthusiasm seems more a new feeling of competency and self-esteem coming from mastery of powerful new skills. This enthusiasm is quickly transferred to senior management, who discover a new, hardworking resource that delivers what it promises on time.

4. **Quantum-leap improvements in productivity are possible.** Making the transition from electric typewriters and calculators to microcomputers is qualitatively different from the previous transition from manual to electric typewriters and back-of-the-envelope mathematics to calculators.

A secretary using an electric typewriter can type a report perhaps 20 percent faster than one with a manual typewriter; a microcomputer can speed up production times 4 to 5 times, or more if many cycles of correction are involved. The time saved is not only that of the secretary, but the much more valuable time of the professional staff no longer needed for multiple cycles of proofreading. Furthermore, unlike applications on mainframe computers, major systems can be developed on microcomputers in a relatively short time. The Government Budget exercise described earlier, including all equipment installation, training, programming, and data entry for the entire Government, took only one year. In addition to saving time, microcomputers allow much greater flexibility in the presentation of data than is possible with manual systems.

Micros initially should be used in any area where significant productivity gains are possible, including word processing. Promoters should plan to purchase enough microcomputers so that productive applications are never held up by equipment shortages. Donors and TA contractors should keep in mind that the cost of a microcomputer work station, including a printer, consumables, and maintenance for a year, is equivalent to the cost of an expatriate advisor or consultant for 1 to 3 weeks.

5. **Many of the benefits of microcomputers will be unexpected.** Microcomputer technology is so far removed from the experience of most African managers that they won't be able to anticipate in advance where it will be most useful. Thus, elaborate feasibility studies preceding the introduction of microcomputers are likely to be a waste of time. The MFP's microcomputers were originally introduced to improve operating efficiency in the Planning Department. However, major targets of opportunity soon arose in the Finance Department. Many of the improved planning systems sought by the Planning Department could only be achieved by parallel improvements in financial management systems, such as the Government Estimates. Since computerization is particularly well-suited to these applications, the management advisor and his staff spent a disproportionate amount of time developing microcomputer-assisted systems for the Financial Secretary. This was fortuitous, since by helping him better understand the promise of the new technology, he was favorably disposed to approve much needed additional microcomputers and technical-support staff for the Ministry.

In retrospect, among the major benefits are the improved working relationships between officers from the Planning and Finance Departments. Although not fully anticipated, the Planning Department was able to take advantage of these benefits because: (1) the Department was not locked into a rigid program for using the equipment; and (2) the Department had planned to have available extra equipment and technical staff to respond to just such targets of opportunity as they arose and still have sufficient resources to cover its internal requirements.

6. **Train more people than needed.** For microcomputers to gain credibility, it is important that local staff take over most operations as quickly as possible. In the MFP, this was done after less than one year. Government staff were able to handle all data entry, all report production, most training, some programming and new system development, and most hardware repairs (e.g., disk read/write errors, blown fuses, loose chips). Local consultants were able to handle all other programming and training. The advisor was needed for "salesmanship," overseas procurement, major systems design, and overall coordination and management.

Up to now, the most enthusiastic users are lower-grade staff without University degrees. There are many reasons. Lower-grade staff quickly realize that microcomputers make their monotonous jobs more interesting. Self-esteem increases from mastering a new skill, and from new opportunities to interact with higher-status officers. Many of them already know how to type, which facilitates mastery of the computer. Many also see suitable moonlighting opportunities, and even new, permanent positions outside government at much higher salaries. For this reason, it is critical to train more people than needed because some microcomputer operators will inevitably find more attractive positions outside of government service.

The most critical manpower problem is the lack of interest on the part of professional staff in mastering the technology. Although some have learned enough to use microcomputers in their own work, few have mastered them to the point where they can manage a major exercise like the Estimates. Indeed, many of the most effective Budget Officers would seem, like some of their counterparts in the developed world, to suffer from "cyperphobia."[19]

7. **Outside technical assistance is critical during the initial stages, yet the operation must not become a foreign one.** Microcomputers were introduced in the MPF by two advisors on two-year contracts. Both were hired to work in other areas, and both concluded that microcomputers would be useful in carrying out that work.

Since the advisors were not microcomputer "experts," they recruited local consultants to do training and programming. The advisors were useful in procuring the equipment, and in persuading senior management to make use of it despite their political reservations. Although the advisors used microcomputers to do their own work, most of the new financial management systems were programmed and operated completely by local consultants and Ministry staff right from the start. Thus, the microcomputer operation was widely perceived as Kenyan—a fact critical in overcoming the political opposition to the new technology.

Because senior management and GCSC were initially so uninformed about microcomputers, the advisors played an essential, interpretive role. They were the bridge between the substantive concerns of senior management and the technical implementation of appropriate systems by the microcomputer consultants and Government programmers. The advisors were effective because they were trusted by senior management for their understanding of the substance of government operations.

8. **When computerizing governmental budgets, begin if possible with the Ministry of Finance.** In Kenya, the first ministry to develop a computerized budgeting system was MALD. Although the system developed was extremely useful to the Ministry, it was of limited value to the MFP because the format was different from the one used by MFP. Now that the MFP has its own system, the software and

hardware are incompatible with those used by MALD. Thus, all MALD data must be entered twice.

Ideally, the MFP system should have been developed first, and then replicated by other ministries. In fact, there had been a Government regulation that all computer activities of other ministries be approved by the GCSC. It was difficult to enforce the regulation when ministries had their own source of donor finance, as was the case at MALD. This turned out to be fortuitous, since success in computerized budgeting at MALD pointed the way for such a system at MFP. However, there was the cost of incompatability, which now is being paid.

9. **Don't underestimate the logistical difficulties of getting equipment in place and keeping it running.** Although much equipment is available in Kenya, it costs 50–500 percent more than similar items in the United States. Furthermore, when a particular spare is needed for a repair, it is rarely available.

Thus, most equipment is purchased in the United States or United Kingdom by a contractor, and air-shipped to Kenya. This system works fine except for the delays of up to six months between sending an order to the United States and taking receipt in Kenya. There are many reasons for this. Contracts may require written bids, and many suppliers are reluctant to do this for relatively small orders. Export licenses are required for shipments valued over US$1000, a process taking 6 to 12 weeks. It is therefore important to have a much larger inventory of spares and backup machines than would be needed elsewhere.

Repair services are also a problem. Some technicians in the private sector begin repairing microcomputers after only a few weeks of training. While not necessarily a problem, some local technicians, either through lack of knowledge or overwhelmed by a rapid growth of their business, have misdiagnosed problems, and made the Ministry's microcomputers worse off than they were before repair. Ministry staff have been trained successfully to repair certain common problems in a few hours.

Summary and Conclusions

Microcomputers are improving the budgeting process and other data processing activities within the MFP by making staff members and technical assistance personnel more productive. Productivity increases are particularly dramatic for clerical and support staff. Both because of improved motivation and the power of the new technology, these staff can now turn out complex reports with impressive speed. These reports are more useful tools for decision makers than those before because they are more timely, have fewer errors, have greater analytical content, and contain exactly what is needed for specific purposes. The new microcomputer based systems allow far more productive use of scarce professional and advisory manpower by reducing the need for repetitive cycles of data entry and proofreading, and manual calculations.

Despite some problems, the Budget Department views the computerization of the 1985/86 Estimates as a stunning success. All future budget documents will be produced using microcomputers.

Although progress has been impressive, two major problems remain:

1. Microcomputers have been mastered only by a handful of senior and mid-level professionals. Since most professionals have not mastered the technology, they cannot properly manage the Ministry's computer operators. Expatriate advisory staff are temporarily filling this void. Such a system will not be long sustainable.
2. Talented, low-grade programmers need to be upgraded to professional status, or they soon will be lured away from the Ministry to more lucrative, higher-status positions in the private sector.

If these manpower issues can be resolved, microcomputers will become increasingly more useful to the Ministry, and will quickly spread throughout the rest of the Government.

Acknowledgments

The author gratefully acknowledges the assistance of Christopher Misaro, David Green, Noel Berge, Charles North, Tom Caruso, and John M. Cohen in writing this chapter. The exercise described was supported in part by the Rural Planning Project, Phase 2, funded under contract with the Harvard Institute for International Development by the United States Agency for International Development, and in part by the Microcomputers for Improving Data Processing Project, funded under contract with Thunder & Associates by the same donor. Most of the microcomputers used were donated by the US-based Kaypro Corporation.

Notes

1. Called the 1985/86 Estimates, the Budget is comprised of the budgets for 28 Ministries and sub–Ministries for the financial year 1 July 1985 to 30 June 1986.

2. In August 1985, after the 1985/86 Budget exercise was completed, the MFP was divided into the Ministry of Finance (MOF) and the Ministry of Planning and National Development (MPND).

3. A set of administrative procedures of the Kenya Government designed to move toward the objective that the 41 administrative districts become the focus of rural development. See Working Party on Government Expenditures, *Report and Recommendations of the Working Party*, 1982, pp. 50–55; Office of the President, *District Focus for Rural Development*, 1984.

4. Most Government computing is carried out on an IBM 370/138, located in the GCSC's headquarters in Herufi House. The GCSC also maintains a Wang 2200 MVP, and IBM 4331 Model 11, and two microcomputers (Comart and Midas). According to a recent report, the GCSC presently has three years of arrears of income tax data awaiting data entry and validation. Migration statistics are one year in arrears, trade statistics lag by four months. The motor vehicle master file hasn't been updated since 1981. Data from the Rural Roads Impact Study are waiting to be entered. Data from the Rural Household Budget Survey (1981/82) are still being cleaned. The Primary and Secondary School Censuses have a two–year backlog. Processing delays are so chronic that "...manually processed information was nearly always prepared more quickly than computer–prepared information...." "Statistical Information Processing Plan– Kenya." (Paper prepared for UNDTCD/UNDP, 1985), p. 2.

5. See Chapters 5 and 7.

6. In recent years, microcomputers have similarly been found appropriate in government settings in Kenya and other countries. See T. J. Cartwright, *Microcomputers—Appropriate Technology for Developing Countries* (Nairobi: Faculty of Environmental Studies, York University working paper in Environmental Studies, Kenya Series, No. 6, April 1984); Stephen B. Harsh and Michael T. Weber, "What do we Know About the Use of Microcomputers in Agriculture and Natural Resource Sectors in Developing Countries" (Paper prepared for Symposium on the Use of Microcomputers for Management and Science, sponsored by Computer and Information Technology Council, Colombo, Sri Lanka, and Board on Science and Technology for International Development, National Research Council, Washington, D.C., November 1984); Carl H. Gotsch, *Applications of Microcomputers in Third*

World Organizations (Chapter 3, this volume); James R. Griesemer, "Microcomputers and Local Government: New Economics and New Opportunities," *Public Administration Review*, January/February 1984, pp. 57–9; see also other articles in *Ibid* on related subjects, and works cited in Harsh and Weber (*op. cit.*) and Brodman, Chapter 7, this volume; there are also less optimistic accounts, such as Schware and Trembour, Chapter 2, this volume.

7. The goal of this eight–year technical assistance and training activity was to increase governmental efficiency and responsiveness by promoting decentralized administration and planning.

8. The Osbornes were acquired because there were already two other Osbornes installed in the Ministry dedicated to industry studies and trade data collection and analysis. Also, CP/M software seemed appropriate for the envisioned tasks, and local dealer support was available, including willingness by the dealer to give the Ministry computers on an approval basis so that the Ministry would have time to evaluate their suitability and arrange for financing. The portable Osborne was also selected so that it could be easily carried to the GCSC to download data from the IBM 370/138 computer. Another advantage of the Osborne running the programs "Wordstar," "Supercalc," and "dBase II" was that a number of computer consultants in Nairobi had experience with these packages running on the Osborne. The Apple was acquired in the hope that Apple format data disks from the MALD could be used on the Apple itself, or transferred using communications software to one of the Osbornes for integration with other data sets. At that time, neither the IBM PC nor any compatibles were supported in Nairobi by any dealer.

9. District Development Plans are prepared by each District every five years, covering the same period as the 5–year National Development Plan. The goals are to provide District–level input into the planning process, and to lay out an operational development strategy and work program for each District consistent with national priorities.

10. The fact that some of the microcomputers are operated by end users rather than data–processing specialists is a major departure from prior Government computing policy. A major problem of a general service organization such as GCSC is that end users have little understanding of the technical capabilities of the equipment, and tend to design surveys and other data processing jobs as if computing capacity were boundless; thus, the backlogs. On the other hand, data processing specialists have a limited understanding of the substantive purpose of any given job; thus, programming and computing time may be used unproductively. Because of the relative simplicity of micro-

computers, such problems can be alleviated. With end users operating their own equipment, they have a clearer notion of its capacity, and of its most productive use. The GCSC will continue to be important, but its role will shift. In the case of microcomputers, GCSC staff will become less involved in data processing *per se*, and more involved as providers of training, technical backup, and other services.

For more details on the structure and responsibilities of the MIS, and on representative microcomputer applications, see Clay G. Wescott, *Microcomputers for Improved Budgeting by the Kenya Government* (Cambridge, Mass.: Harvard Institute for International Development, Development Discussion Paper, No. 227, 1986).

11. MFP, *Compendium on Development Cooperation* (Nairobi: Government Press, 1985).

12. The previous section has described some of the improvements in Planning and Financial Management which were assisted by microcomputers. For a more complete listing, see Wescott, *op. cit.*. See also Clay G. Wescott, *Microcomputers for Improved Project Monitoring in Kenya* (Cambridge, Mass.: Harvard Institute for International Development Discussion Paper, forthcoming) and Rural Planning Project *District Microcomputer Experiment* (Nairobi: Ministry of Planning and National Development, 1985).

13. For comparative perspective on budgetary reform in other developing countries, see Doh Joon-Chien, "Modernizing the Budget System: the Malaysian Experience," *Public Administration and Development*, Vol. 1, 1981 pp. 291–305; D. J. Hardman, "Public Financial Administration of Microstates: South Pacific Forum," *Public Administration and Development*, Vol. 4, 1984, pp. 141–54.

To understand fully the program of work, it is useful to be familiar with the structure of the MFP and of the Estimates. The top civil servant in the MFP, as in any Ministry, is the Permanent Secretary, who reports to the Minister and Assistant Ministers. Under the Permanent Secretary is the Financial Secretary and Director of Planning, who head up the Finance and Planning Departments, respectively. Under the Financial Secretary are, among others, the Director of Budget and the Director, GCSC. The Chief Estimates Officer reports to the Director of Budget, and takes operational leadership over the Estimates process. The Microcomputer Section, established during the period under review, comes under the GCSC. The Management Information Section and Management Advisor come under the Rural Planning Division, and in turn, the Planning Department.

The Annual Estimates are produced each year by the Budget Department, based on submissions by Ministries. The Estimates are printed in two volumes: the Recurrent Estimates, covering salaries, operating and maintenance expenditures, and the Development Estimates, covering capital expenditures. Each volume is divided into votes, or financial allocations for a particular Ministry or sub–Ministry. Each vote is divided into hierarchical categories of subvotes, heads, subheads, and items. In addition, the Recurrent Estimates contains a listing under each vote of all approved posts, the numbers of staff in each post, and the estimated expenditure for each post. The Recurrent Estimates also include a listing of the numbers of vehicles under each vote by Department and vehicle type, and accounts for the Consolidated Fund (debt service, pensions, etc.). The Development Estimates also contain a listing of all loans and grants available to Government, and anticipated funds available under each for the 1985/86 fiscal year. The Estimates are published in considerable detail. For example, the 1985/86 Estimates contained details for two fiscal years (1984/85 and 1985/86) on 28 votes, and over 26,000 budget items and posts, and covered over 2000 printed pages. In addition, the Budget Department started in fiscal year 1984/85 preparing two additional volumes giving allocations by District for all items used to finance District–level costs. For sample formats of budget documents, see Wescott, *op. cit*.

By law, the two main volumes must be submitted by the Minister to Parliament by the third week of June.

14. Another problem was the Recurrent Estimates for the Ministry of Transport and Communications. The first draft of this vote was not received by the computer until 2 May. In original typewritten form, it was nearly 200 pages long. It was structured completely differently from the previous year's estimates; nearly every head was divided into district allocations. Because of the new structure, the computer section could not simply add the present year's Estimates to the existing file. The entire budget had to be re–entered, a process taking one week. On the day this had been completed, one of the Budget Officers complained at the daily staff meeting with the Financial Secretary that there were serious delays in the computer section. The Management Advisor noted the problems described here. The Financial Secretary said that the Ministry could not change the structure of its vote without clearance from the Head of Accountancy Services. The Budget Officer would have to redo the vote; the computer section would have to re–enter the data using the old format.

15. The Microcomputer Advisor assisted the Head of the Microcomputer Section with programming and equipment repairs, and provided overall leadership, with specific responsibilities including preparing Terms of Reference, hiring and supervising consultants, handling

procurements and outside repairs funded by RPP2, liaison with the Financial Secretary on strategic issues, and representing the microcomputer section in meetings with the Budget Department. The microcomputer consultant from Thunder & Associates had only a minor involvement with Budget work during this period.

16. Fortunately, there were no major equipment failures during the sixweek period. One LQ–1500 printer frequently seized up during the last two weeks, apparently from overheating. This printer also developed an annoying habit of printing random, incorrect characters in the middle of a file. Both LQ–1500 printers stopped working as soon as the budget was completed. Most of the Kaypros ended up with at least one broken key, but this did not seriously hinder work. The FX–100 printers stopped working frequently during the period. In most cases, they were quickly repaired by microcomputer staff. In one case, the machine was repaired by a local agent. Fuses in plugs blew out frequently, but they were readily repaired.

Media failure was also surprisingly rare during this period. There were at least two cases where current dBase II files lost all data. Both cases seemed the result of using IBM PCs to print the text files. The IBM PCs had to be used since all other equipment was in use for data entry. dBase II files on the Kaypro–format disks all used file names including colons (:), which the IBM does not recognize. Problems occurred when the IBM PC, running the Uniform Program Set to Kaypro format, was used to make corrections to the text files before printing. The text files (whose names didn't include a colon) were saved correctly, but the dBase II files could be damaged. In both cases, reasonably current backups were available, so little time was lost re–entering data.

17. Report and Recommendations, (*op. cit.*).

18. For example, one of the Ministry's top microcomputer programmers is a Job–Group D Clerk with no college education. He has worked for the last year virtually every weekend and most nights. In addition to dBase II and Supercalc programming, he does training, data entry, equipment repair—whatever needs to be done. As he put it one day: "...before, I had a job; now I have a career." Whether his long–term career prospects are with the public or private sector remains to be seen. In any case, his mastery of this new technology has led to a fundamental change in his attitude toward work: from a compartmentalized, time–serving mentality into a new sense of pride in professional growth and achievement. Although the Ministry has not yet found a way to promote him, it has, in the time–honored tradition,

subsidized his participation in a local, degree-granting program in computer science, using RPP2 funds.

19. "...an extreme anxiety produced by dealing with a high-technology computer system." Marie R. Stempinski, "Automation Can Cause Computer Anxiety," *Government Computer News*, Vol. 4, No. 15, August 16, 1985, p. 66.

7

Key Management Factors Determining the Impact of Microcomputers on Decision-Making in the Governments of Developing Countries

Janice Z. Brodman

Introduction

The rate of microcomputer adoption by the governments of developing countries is increasing rapidly. A central goal of microcomputer use is to improve access to information, thereby producing more informed, better-reasoned decisions. For most such microcomputer adoption efforts, attention is focused largely on technical issues, i.e., hardware and software choices, and training in microcomputer operations. Of course, these "technical" requirements must be met. Hardware and software must be obtained, microcomputer operators trained, and sources for service, parts, and reliable power be arranged. Yet there is a critical aspect of microcomputer adoption which generally receives far less attention than its importance warrants. The evidence from case studies of microcomputer use in developing countries suggests that key determinants of microcomputer impact are associated

An earlier version first appeared as Development Discussion Paper No. 231 of the Harvard Institute for International Development, Cambridge, Massachusetts, June 1986, "Microcomputers to Improve Decision-Making in Third World Governments."

with management systems and structures rather than hardware and software.[1] These observations are supported by case studies of microcomputer adoption in the West.[2] They also conform to findings in the wider literature on organizational change through adoption of new technologies.[3]

The Importance of Management Factors

Four management factors emerge as principal determinants of the extent to which microcomputer use improves the quality of decision-making. The first factor concerns leadership messages. Highly visible leadership support has been cited in the literature as critical to the adoption of innovation in an organization.[4] Explicit leadership support is important not only to ensure requisite resources, but also to legitimize the innovation, to build a common vision of change, and to convince people that the change is real.

Studies of microcomputer adoption in developing country government agencies suggest that clear, consistent, positive messages from bureaucratic and/or public leaders are key to establishing the use of microcomputers for improved decision-making in the organization. Those messages include not only explicit statements but also implicit signals. If, for example, leaders ignore work performed with microcomputers, staff may infer that the leadership considers microcomputers to have little value.

The second factor involves requisite training. The evidence suggests that training in the **use of the information** provided by the microcomputers is as important as training in the operation of the microcomputers. Even if the microcomputer system only provides the same information as was produced manually, system designers must understand both the principles of microcomputer operations and the way the information is to be used.

Furthermore, microcomputers often are used to produce information that was not previously available. In such cases, those using information from the microcomputer system must be trained in the analysis of the information, and the application of information to decision-making. Other types of collateral training may also be necessary if microcomputers are to help improve decision-making. Professional staff, for example, often lack skills in "packaging" technical information so that it is readily accessible to decision-makers. Managers may need training in new management techniques if they are to utilize effectively the microcomputer information system.

The third factor concerns the incentive system. Many studies of microcomputer adoption in the West now routinely investigate the impact of "orgware" (human and organizational factors). One of the most important "orgware" factors is the incentive–sanction system. Experience in developing countries makes it clear that the incentive-sanction system is equally critical in those contexts. The incentive system must support microcomputer use if people are to invest the time and effort necessary to learn to use the microcomputers and the microcomputer output.

Incentives can be tangible or intangible, formal or informal. Positive feedback and recognition from senior officers, for example, can provide powerful incentives for junior staff to learn to use microcomputers. Microcomputers themselves provide certain inherent incentives such as making tasks easier and more interesting.[5] There are also external incentives such as the perception that microcomputer skills may expand employment opportunities outside the organization.

The fourth factor is the power structure, which strongly influences the extent to which information provided by microcomputers affects decision–making.[6] In most cases of microcomputer adoption, there is likely to be some conflict among decision makers about the value of microcomputers. Although acceptance of microcomputers need not be unanimous among decision makers, those with power over resources critical to effective microcomputer use must support adoption.

These four management factors appear to be particularly important when microcomputers are intended to bring about certain kinds of change. Although microcomputer use often has multiple objectives, for the sake of clarity, we can distinguish between two types of goals. One involves a "technical" solution, i.e., the goal is only to achieve greater accuracy and faster processing of the same information that is already being used. The second involves a "comprehensive" solution, i.e., the objective is to induce new ways of using information, or use of new kinds of information.

If only a "technical" solution is sought, there is a high probability that microcomputer adoption can make a significant contribution. The major components of success will be introduction of the microcomputers and appropriate software, and training staff in microcomputer operations—relatively straightforward tasks. The use of microcomputers by the Ministry of Agriculture and Livestock Development (MALD), Kenya, to distribute Authorities to Incur Expenditure (discussed in Chapter 5) is a good example of a limited, "technical" goal.

If, however, a "comprehensive" solution is needed, merely introducing microcomputers and software, and training staff to operate them, will not be enough. Successfully changing the way information is used will depend largely on the attention to the four management factors cited earlier. This chapter is concerned with these types of applications.

The chapter examines three cases of microcomputer adoption to show how the four management factors affect the achievement of "comprehensive" goals. The first case is MALD which received its first microcomputer in late 1981 (Chapter 5). By 1985, the Ministry had seven microcomputers. The second case, the Ministry of Finance and Planning (MFP), Kenya, adopted five microcomputers in the spring of 1983. The Ministry had more than 25 microcomputers by late 1985 (Chapter 6). The third case is the Ministry of Finance (MOF), Indonesia. MOF adopted a few microcomputers before 1984, but the machines had limited use and negligible impact. In the summer of 1984, the Ministry adopted 200 microcomputers as part of a major effort to reform certain Ministry processes.

Although the three ministries have used microcomputers for a relatively short period, they are on the cutting edge of microcomputer use in the governments of developing countries. Their experiences therefore provide valuable insights into the potential benefits of and obstacles to effective microcomputer use.

Examination of the Factors in the Context of the Cases

Kenya: The Ministry of Agriculture and Livestock Development; Finance Division

The finance division of MALD has developed a large number of microcomputer applications since Technical Assistance Pool (TAP)[7] advisors first introduced the machines in 1981. One of the most important applications, expenditure monitoring, is representative of the division's experience in pursuing "comprehensive" goals. It illustrates the critical role that management factors played in a broad range of applications.

Expenditure Monitoring. Before MALD adopted microcomputers, expenditure information from the field was processed too slowly to be of use for monitoring. Moreover, it was replete with errors. Microcomputers were intended to speed up access to expenditure information and to make the information more accurate. They were also to help support the government's "district focus" policy by allowing financial information to be disaggregated to the district level, a task nearly impossible without some sort of computer capability.

A related objective was to improve periodic monitoring to allow intervention when spending stations were over or underspending their allocations. Using hand calculators, the finance officers in charge of monitoring could do little more than simply aggregate figures for year-end expenditure reports.

An additional goal was to induce finance officers to utilize expenditure information more effectively in developing budget allocations. Using manual methods, finance officers had been unable to get accurate expenditure information throughout the year. Consequently, they made allocation decisions with little regard for projects' previous expenditure record. The result was occasionally blatant misallocation of funds.

Progress toward these goals was varied. The first goal, improving the accessibility of information, was essentially a "technical" goal. It was achieved quite successfully. TAP advisors performed the difficult task of designing the system to reflect the capabilities of the microcomputers and the needs of the finance officers. TAP then trained a small number of MALD clerks to program the microcomputers and produce required reports. With the use of the microcomputer system, expenditure reports were produced more quickly and accurately, with greater flexibility to enhance the accessibility of information. The greatest threat to the system was a lack of personnel continuity, since the formal incentive system did not reward clerks who learned to program.

Efforts toward the goal of improving periodic monitoring had a more mixed result. On the one hand, the monthly expenditure reports enabled finance officers to conduct monitoring throughout the year. As a result, they could intervene early when a particular spending station was substantially over– or underspending its budget. On the other hand, this improvement in monitoring had virtually no impact on project spending or on project evaluation. Although spending stations were notified of unacceptable spending levels, there were no follow–up mechanisms to enforce conformity with expenditure guidelines. Nor was there regular analysis of the causes of over- and under-spending.

Factors under management control produced this situation. First, those who conducted monitoring lacked the skills necessary to analyze the information the microcomputers make available. Second, those responsible for monitoring did not have the power to enforce their warnings to the spending stations. Third, those at headquarters who did have the power to enforce rates upon spending stations had no incentive for doing so.

There was even less progress toward the goal of inducing finance officers to improve the application of information in their budget allocation decisions. Although finance officers valued the expenditure monitoring reports, they used the information only for routine reports and to respond to requests for information from above. They neither analyzed the financial information, nor used it in making allocation decisions.

Four management factors obstructed improved use of information for decision–making among finance officers. One factor was leadership messages. Although the finance officers quickly came to view the expenditure monitoring reports as "essential," they doubted the value which senior management placed upon microcomputer skills. Despite the division head's strong, explicit support to microcomputer use, only clerks and secretaries were assigned to learn to operate the machines. Finance officers inferred from those actions that top management in the division considered microcomputer operations appropriate for clerks, not professionals. The finance officers therefore resisted learning even the principles of microcomputer use and its application to decision–making. Only after higher echelons in government had become visibly interested in microcomputers did a MALD finance officer become interested in learning to operate the microcomputers.[8]

The second obstructing factor involved training in the use of the information provided by the microcomputer system. Many finance officers had no academic background in finance or related subjects. There was no regular ministerial program to train finance officers in financial analysis skills (although TAP advisors were providing such training). As a result, although the microcomputers made information more accessible to finance officers, most officers were not equipped to use that information to improve their decision–making.

Another consequence of the limited analytic skills in the division was that the top echelon in MALD could not rely on those below them to provide useful summary analyses of financial information. Potential contributions to decision–making at the top were therefore impeded by the limited training of senior and middle level finance officers. To the extent that microcomputers contributed to improved decision–making at the top, it was due primarily to analysis provided by advisors.

The third factor inhibiting a positive impact on decision-making was a lack of incentives for finance officers to analyze and apply financial information, whether it came from the microcomputer or elsewhere. Nor were there any sanctions for not doing so. Finance officers were therefore reluctant to invest the considerable time and effort necessary to learn to use microcomputers for analysis.

Finally, finance officers were under pressure in many cases to make financial decisions on bases other than financial information. In a context where there were no countervailing incentives or sanctions that encouraged information-based decision-making, finance officers generally responded to the same factors in the power structure as they had done in the past.

Nevertheless, use of microcomputers in the finance division supported some changes in the way senior finance officers used information. The improvements were directly related to training and incentive systems that were, in a sense, superimposed upon the Ministry systems by TAP advisors. Microcomputers contributed significantly, for example, to the training in financial analysis that TAP advisors provided to senior finance officers. Advisors used the microcomputers to perform rapid transformations of the data to highlight important issues, and to format information in ways that clarified its implications. Thus, while microcomputers alone did not improve use of information in decision-making, they proved a valuable tool in a management training effort to do so.

The advisors also offered incentives that encouraged finance officers to learn to utilize information better for decision-making. Particularly effective among these incentives was the opportunity to go abroad for further training. TAP also used intangible incentives. They provided regular, positive feedback, for example, to officers who made an effort to analyze financial information.

Incentives external to the Ministry system also began to operate. Finance officers began to realize, for example, that microcomputer skills could expand their employment opportunities outside the Ministry. Their interest in learning to operate microcomputers increased concomitantly.

Kenya: The Ministry of Agriculture and Livestock Development: Planning Division

MALD's planning division adopted microcomputers in 1983, after microcomputer use in MALD finance had shown growing success. Two of the most important microcomputer applications in the planning division were word processing to produce the projects registry and statistical packages for planning analysis. Although use of microcomputers to produce the projects registry began after use for analysis, it was adopted much more quickly and widely among planning officers. The divergent responses by planning officers to the two kinds of applications reflect the influence of the four key management factors.

Conducting Statistical Analysis. A central reason for TAP's introduction of microcomputers in the planning division was to encourage planning officers to conduct quantitative analysis. Most planning officers acted as administrators rather than planners. TAP hoped that some planning officers would use the microcomputers, providing a model of the kinds of analysis which the planning division could perform.

The response from the planning officers was limited. A few junior officers quickly learned to use statistical packages or spreadsheets. They were largely self-taught, referring to the manuals and turning to TAP advisors when they could not resolve a problem alone. These few planning officers came to use microcomputers regularly for analysis. They worked with TAP advisors to design research models, and to analyze and interpret the results. In addition, a small number of more senior officers voiced interest in microcomputer use.

Most planning officers, however, showed no interest in using microcomputers for analysis. To the limited extent that they did any quantitative analysis, they used a calculator and drew charts and graphs by hand. One important reason for their disinterest was a lack of analytic skills necessary for effective use of microcomputers for planning analysis. The Ministry had no training system to strengthen those skills among planning officers (although, again, TAP advisors were providing such training, largely on an individual basis).

There were also, however, planning officers with analytic skills who ignored opportunities to learn to use microcomputers for analysis. The apparent contradiction reflects the influence of leadership messages, incentives within the division, and a sense of powerlessness.

Senior officers in the planning division, by their lack of enthusiasm, sent negative signals about using microcomputers for analysis. Said one junior officer, "I know the benefits of using the microcomputer, but the senior people feel it's a waste of time. You can do it all and no one reads it."

Some planning officers even suspected that if they used quantitative analysis, senior officers might perceive them to be "showing off." They feared they might alienate their superior officers and reduce their chances of promotion. Many junior planning officers also said that when they were asked to prepare reports for the senior levels, they were given so little time to respond that it was impossible to carry out analysis. Senior officers themselves had little confidence in their own officers' ability to provide the analysis they needed.

Aggravating this situation was a lack of formal incentives for the planning officers to learn to use the microcomputers, or to do quantitative analysis generally. Promotion prospects for economists in MALD were very poor, regardless of the quality of their work. And, as one advisor pointed out, planning officers were not rewarded for doing something right, but were penalized for doing something "wrong." As a result, planning officers avoided attempting anything new. Their risk aversion extended to microcomputers. Many felt it was not worthwhile attempting to learn to use microcomputers for analysis, and it might even be dangerous to do so.

Finally, the planning officers felt that they were cut off from those with policy-making power. There was little communication between planning officers and the policy-making administration division. As a result, planning officers had little interest in doing the kinds of analysis that top administrators might find useful.

The planning officers who did use microcomputers for analysis were responding almost entirely to training and incentives provided by TAP advisors. The advisors worked directly with the officers to encourage analytic work and to help them develop their skills. The microcomputers' intrinsic characteristics facilitated these efforts. Microcomputers were, as one young planning officer said, "more fun and easier to use" than a calculator. Thus, while microcomputers may not have been essential—much of the quantitative work could be done with a programmable calculator—they reduced the drudgery, encouraging a form of analysis that might not otherwise be done.

The kinds of incentives TAP advisors offered the finance officers, e.g., opportunities to study abroad, were also important in encouraging planning officers to learn to use microcomputers for analysis. One influential intangible incentive was a series of seminars arranged by the advisors and a few middle level officers to provide a forum in which planning officers could present their work to their colleagues.

These "external" incentives, and the influence of the early users, began to stimulate wider interest in microcomputer training among junior planning officers. Those who understood how to use microcomputers for analysis have began teaching other junior officers. They did so by taking the role of the advisors vis-a-vis their fellow officers: helping them design the research model, demonstrating how to enter the data, and assisting in analyzing the results.

Producing the Projects Registry. The projects registry already existed in the planning division when microcomputers were introduced. It was extremely difficult to keep up to date, however, and the records were poorly maintained. Shortly after microcomputers were adopted, top ministry officials began putting pressure on the planning division to provide better information about ministry projects. To enable the planning division to satisfy requests from senior ministry officers, TAP encouraged the use of word processing to prepare the projects registry.

Initially, planning officers responded with indifference, and most of the registry production was carried out by secretaries. To some extent, the planning officers' reluctance to use the microcomputers was due to lack of typing skills, which made learning word processing difficult. Far more important, however, were the same management factors that had affected the use of microcomputers for analysis—neutral or negative leadership messages, a lack of incentives, a lack of power. These factors discouraged planning officers from investing time in learning to use word processing. To many, microcomputers seemed merely "the advisors' project."

Nevertheless, the use of word processing was highly successful in achieving its "technical" goals, facilitating preparation and proofreading, and permitting monthly revisions of the lengthy registry. The new system began to capture planning officers' interest. By the end of the first year, the registry had, more than any other form of microcomputer use, involved planning officers in key activities.

Management factors were again key to the planning officers' response. First, it became clear to the planning officers that microcomputers made it easier to provide accurate project information to senior officers. Top officials' increasingly explicit recognition of the role of microcomputers further stimulated planning officers' interest in learning to use the machines.

Also contributing to the planning officers' favorable response was the fact that they could use information from the registry in the same ways as they had earlier. Use of microcomputers—either directly or through secretaries—merely gave them faster, easier access to the information, and greater confidence in its accuracy.

Thus, planning officers responded far more rapidly and positively to using microcomputers for the projects registry than for planning analysis. Pressure from senior MALD officers for faster access to more accurate project information created incentives for planning officers to use the machines. Positive leadership messages about the value of microcomputers for the registry added further encouragement. There was even an emerging view among planning officers that they might gain some influence over policy through their ability to provide the information senior officers demanded. Finally, planning officers could benefit from using the microcomputers without having to learn new ways to use information.

These conditions contrast sharply with those related to using microcomputers for analysis. In the latter area, there was little interest expressed by senior officers, there were no incentives within MALD for conducting analysis, and, in most cases, planning officers did not have the analytic skills necessary for using microcomputers for analysis.

Kenya: The Ministry of Finance and Planning

Microcomputers were introduced into the Ministry of Finance and Planning by the Rural Planning Project (RPP).[9] Among the microcomputer applications in the Ministry, two of the most important are financial and planning analysis, and budget production. Use of microcomputers for quantitative analysis generated little interest among finance and planning officers.[10] The use of microcomputers for budget production had notable success in achieving some important goals; yet it also evoked considerable resistance from some budget officers. Again, the differential response by the professional officers to the alternative microcomputer applications was associated with the four key management factors.

Financial and Planning Analysis. The experience of microcomputer use for analysis in MFP is highly reminiscent of that in MALD. RPP advisors used the microcomputers to provide better technical support for government policy-makers. Yet the benefits to decision-making were largely confined to officers at the top level of the ministry.

During the first year of microcomputer use, contradictory leadership messages inhibited microcomputer use for analysis among middle level officers. During that period, top MFP leaders gave only tacit support for microcomputer use, by accepting increasing numbers of microcomputers into the Ministry and requesting reports prepared with microcomputers. At the same time, other senior officers in MFP, including those in the microcomputer services center, were highly critical of microcomputer adoption. In addition, concern was expressed at the highest levels of government about possible labor displacing effects of word processing.

The finance and planning officers responded to these mixed signals from their senior officers by questioning the value of learning to operate the microcomputers. They wondered, during the first year of microcomputer adoption, whether microcomputers would be retained in the Ministry after the RPP advisors went home.

During the second year of microcomputer use, the official government position toward microcomputers became more positive. Senior MFP officials began to make a greater commitment to them. As they did, finance and planning officers' enthusiasm for using the machines increased. Interest in using microcomputers for file maintenance and monitoring became more widespread. Involvement in using the machines for quantitative analysis, however, developed much more slowly.

A second obstacle to officers' use of microcomputers for quantitative analysis related to limited analytic skills. Although many finance and planning officers received formal training in quantitative analysis, they were often absorbed with administrative tasks and rarely applied those skills. Disuse eroded their ability to conduct the kind of quantitative analysis for which microcomputers are most appropriate.

The third impediment was a lack of formal incentives for finance or planning officers to conduct analysis. Planning officers, particularly, often perceived it unlikely that they would be rewarded for using microcomputers to conduct analysis. Although microcomputer skills might enhance job opportunities outside the Ministry, most planning officers perceived such jobs to be entirely beyond their reach.[11]

The power structure represented the fourth obstacle to the use of microcomputers for analysis. As in the MALD, many important decisions in the Ministry were made on bases other than financial data. To some finance and planning officers, the balance of power seemed to weigh against the use of analysis as the basis for making such decisions, particularly decisions about resource allocation. Without such countervailing forces as formal incentives, officers tended to disregard information provided by microcomputers, and make decisions as they were accustomed to doing in the past.

Budget Production. MFP leadership introduced microcomputers into the budget production process to achieve two kinds of objectives.[12] One goal was to facilitate the physical production of the budget by improving the speed and accuracy of calculations, permitting corrections to be made more easily, and eliminating many of the routine proofreading and retyping tasks. In previous years, the budget was inevitably delivered at least two weeks late to Parliament. The leadership hoped microcomputers would help them to deliver the budget on time.

A second objective was to improve the substance of budget allocations. Time constraints usually forced budget officers to cut line ministry budget submissions with little information from the ministries about project priorities and requisite resources. RPP advisors and MFP leadership expected the microcomputer system to enable Budget Supply Officers to seek more input from the line ministries and improve their analysis of the ministries' budget submissions.

The use of microcomputers in the budget production process of 1985 clearly attained the first goal. MFP completed the budget two weeks before the deadline—a striking success. The achievement was acknowledged by the Parliament and the Office of the President.

Yet not all MFP finance and Budget Supply Officers were enthusiastic about the use of microcomputers. There was strong resistance to the microcomputer system from many Budget Supply Officers who were reluctant to change the way they had prepared the budget in the past. To some extent, the budget officers' criticism of the microcomputers justifiably pointed out the frequency with which errors went uncorrected during the budget production process. Those problems were finally rectified by having advisors supervise quality control.

Far more important to the Budget Supply Officers' response, however, were issues of power and incentives. For many officers, the microcomputer system presented a threat. Previously, Budget Supply Officers held a great deal of autonomous authority in the budget process. When the microcomputers were introduced into budget production, several new professionals accompanied them. Not only did these professionals understand the microcomputers far better than did the Budget Supply Officers, they also sharply criticized the fundamental budget preparation process. The budget officers reacted to this assault on their authority with strong resistance to the entire microcomputer system.

In addition, Budget Supply Officers had no incentive for supporting use of microcomputers in the production process. Although top ministry officials were visibly committed to using microcomputers to produce the budget, they had excluded the Budget Supply Officers from the new system insofar as possible. The message inferred by the budget officers was that the microcomputer system would ease their responsibilities for production but simultaneously would usurp their authority.

These conditions strongly impeded attainment of the second goal, improving the substance of budget allocation decisions. Early completion of the budget gave Budget Supply Officers considerably more time to analyze the finance issues reflected in the budget. Although they did some routine checks to ensure that line ministry submissions were within their ceilings, they conducted virtually no substantive analysis. Nor did they spend more time conferring with the line ministries. Readier access to information merely meant that Budget Supply Officers more quickly made the same decisions they would have made without the use of the microcomputer system.

Limited financial analysis skills among some Budget Supply Officers contributed to the problem. In addition, the budget officers were expected to use entirely new, more participatory management skills. Yet they were given no explanation of the new management approach expected of them. They required training in the kind of participative management style that the leadership wanted them to adopt.

The Budget Supply Officers also needed some incentive for changing (or sanctions for not doing so). Instead, informal incentives encouraged them to resist the new system, which threatened their power. The situation also discouraged any impulse the budget officers might have had to increase the involvement of the line ministries in budget allocation decisions, since that would further erode their authority. Thus, the Budget Supply Officers responded the way most managers react when their power appears under attack: they dug in and protected their authority from any further encroachment.

Indonesia: The Ministry of Finance

The introduction of microcomputers into the MOF, Indonesia, contrasted in several ways with microcomputer adoption in MALD and MFP, Kenya. First, the initiative to adopt microcomputers was made by the head of MOF's computer center (PAIK), rather than at the suggestion of advisors. Second, MOF adopted 200 microcomputers, a much larger number than did the Kenya ministries. Third, microcomputers were introduced as an integral part of the larger Ministry computer system. Fourth, MOF initially placed the microcomputers in district offices rather than only in headquarters. Fifth, although one goal of adoption was to improve the accuracy and speed of information processing, another explicit goal was to alter the distribution of decision-making power between headquarters and the district level of the Tax and the Customs Directorates.

Despite these differences, the four management factors—leadership messages, training in the use of information, the incentive system, and the power structure—were as important to the successful attainment of the goals of microcomputer adoption in Indonesia as in Kenya. This discussion focuses on the Tax Directorate, where microcomputers were introduced more widely than in the Customs Directorate. It is important to note, however, that the key factors discussed below have been equally influential in the Customs Directorate.

Improving Files and Reducing Irregularities at the District Level. One objective of introducing microcomputers into the Tax Directorate was to maintain more accurate and efficient files on taxpayers. The microcomputers were also part of a new system of tax reporting that was designed to reduce irregularities in tax reporting and collection at the district level. PAIK expected that the microcomputers might eventually pay for themselves by helping to increase income tax compliance and standardize tax collection procedures at the district level.

Under the previous tax collection system, district tax officers went into the field to assess and collect taxes. The tax payments were recorded manually at the district tax offices, which retained the details of taxpayer information. Aggregate statistics were prepared manually and sent to the provincial level, along with the taxes that had been collected.

Under the new system, there was no initial contact between district tax officers and taxpayers. Taxpayers voluntarily sent their tax forms and payment to the district office. A district officer immediately entered the information from the tax form into a microcomputer. The data were then entered a second time by another officer and the two records automatically checked for consistency by the microcomputer system. The districts sent a tape of the taxpayer information to the regional office each week.

At the district level, microcomputers were to be used only to prepare information to be sent to the regional office, and to maintain files on tax information. At the regional offices and at headquarters, a minicomputer/mainframe system kept records, monitored tax collection, automatically audited tax returns, and analyzed trends in individual and district tax payments. The district offices were made aware of the several cross–checks performed on the tax information at the regional and central offices.

Although the system had been in place for only a short period at the time of this study, there appeared to have been considerable progress toward headquarters' goals. Tax information seemed to be more accurate and readily available. Early difficulties with uncorrected data errors were curbed through additional training and experience at the district level. In many districts, tax revenues increased after the microcomputer system was installed. The greatest technical problem related to transferring information from district to regional offices. Mailing streaming tapes was more secure than the original plan to mail computer disks. Yet issues of information security continued to demand attention.

The four key management factors played a central role in MOF's experience. There was strong, explicit support for microcomputer adoption from top political leadership as well as from leaders within MOF. These messages contributed to relatively rapid, widespread acceptance of microcomputers in MOF. Although there was continuing negotiation over such issues as how the microcomputers would be used and who had control over them, clear leadership messages contributed significantly to overcoming resistance to the new system.

Training in microcomputer operations was relatively simple due to easily understood, menu–driven programs in Indonesian, which PAIK designed for use in the districts. PAIK staff, along with staff from International Business Machines (from whom MOF bought the microcomputers), introduced the machines with easy first steps, including computer games. As people became more comfortable operating the microcomputers, they were introduced to the programs related to their work.

In this early stage of adoption, PAIK made the conscious decision to use the district microcomputers only to replicate the output that was being produced (or was supposed to have been produced) manually. Aside from the maintenance of taxpayer files and preparation of reports to be submitted to the regional office, the district office microcomputers were used only for limited administrative work. Consequently, training in the use of the information provided by the microcomputers had not yet become necessary. In fact, under the new system, the district offices did less information processing, since they merely had to forward taxpayer information to the regional office rather than use the information to assess taxes.

A more difficult training problem for MOF related to developing the skills required for designing the information system. A major challenge was bridging the gap between "substantive" knowledge about tax information and "technical" knowledge about the principles of microcomputer operations. As of mid–1985, the Kenyan ministries had not faced this issue because the advisors provided the expertise that integrated the substantive and the technical knowledge necessary to design the information systems.

For MOF, however, integrating substantive and technical knowledge posed a serious problem. To bridge the gap, MOF established teams of tax officers and PAIK officers. Through collaborative work, tax officers gained better understanding of the principles of microcomputer operations and PAIK staff developed better comprehension of tax issues. Yet certain problems persisted. PAIK staff still made some decisions that did not accurately reflect the substantive information needs. Tax staff often wanted simply to transfer all information from paper to microcomputers, in ways that did not exploit the advantages of microcomputers. A sustained effort to develop cooperation between the two groups continued to be necessary to make headway in resolving the problems.

MOF also planned to use microcomputers for financial analysis. PAIK expected to establish a systematic program to train regional and headquarters staff to conduct such analysis. Although the viability of such training programs of course could not be predicted by the study, PAIK's recognition of the need for such training is worthy of note.

PAIK also has addressed explicitly the need for incentives to induce staff to learn to use the microcomputers. They anticipated future extra compensation for Ministry staff who learned to operate the microcomputers. Meanwhile, the incentives included training in microcomputer skills. MOF staff perceived the training to be valuable because leadership messages commended such training, and because they believed such training improved employment opportunities in the private sector.

The most significant difficulty in establishing microcomputer use related to the power structure. Although some district offices responded positively, top staff of many district offices were unhappy with the new system. They pronounced the system unworkable, and declared that microcomputers could not handle the information needs of their offices. They complained bitterly about every mistake in microcomputer-produced reports. They made clear their general disgruntlement with having the machines placed in their offices. They attempted to withhold the resources over which they had some power, notably local staff cooperation in establishing the new system. Yet the power of headquarters, and the incentive for district officers to learn to use the new, high-status machines, appeared to have been able to offset the influence of the district leadership.

To a significant extent, the resistance of district office leadership was not to the microcomputers *per se*, but to the new system which transferred their power to headquarters. For some, this was a reaction to the perceived threat to opportunities for private gain. But for many, it was a response to the loss of decision-making authority to headquarters, which they often perceived to be out of touch with local conditions.

There was no attempt on the part of the MOF leaders to design incentives that might win over the district heads. Given the real losses in power that district heads were experiencing, it would have been difficult to provide sufficiently strong incentives to gain their support for the system. The resolution of the conflict will depend largely on the relative power of headquarters and the district leadership in determining the *de facto*, if not *de jure*, processes by which local offices are run.

Conclusions

Management systems and structures, rather than technical issues of hardware, software, and computer skills, emerged as key to determining the response to microcomputer adoption in the three ministries under examination. Primary among those management factors were: (1) the role of government leadership; (2) training in the use of information in decision-making, (3) the ministry incentive system, and (4) the power structure.

Leadership messages were important in convincing staff that microcomputer use was supported by the leadership, and that microcomputers were to be a permanent part of ministry operations. Where those messages were inconsistent or nebulous, staff tended to infer that senior officers considered microcomputers unimportant to their work. In some cases, professional staff regarded mixed messages as a signal that senior staff did not want them to use the microcomputers. The professional staff responded by ignoring opportunities to learn to operate the microcomputers. Where leadership messages were strong, positive, and consistent, acceptance of the microcomputers was established more quickly and widely.

The cases also illustrate the importance of providing training in the **use of information**. The relative importance of such training depends on the goals of microcomputer use. Some of the ministries' objectives involved providing the same information as was produced previously, only faster, more easily, and accurately. In those cases, training in microcomputer skills proved to be sufficient.

Often, however, the ministries set more "comprehensive" goals, which involved using information in new ways in order to improve decision–making. Achievement of those goals required that staff be trained in the use of information. If the professional officers do not receive training in the use of information, it appears unlikely that microcomputers will have a positive impact on the quality of decision–making. It is also important to note that microcomputers themselves supported advisors' efforts to conduct such training. By providing information more easily, quickly, and in more accessible formats, microcomputers can serve as valuable tools in training officers to analyze information.

A substantial literature documents the importance of incentives in inducing acceptance of organizational and technological change. The cases examined here demonstrate some incentives inherent in microcomputer use, e.g., using microcomputers is faster, easier, and/or more interesting than manual methods. They also make clear the importance of the organization's incentive system in determining the impact of microcomputer use.

In MALD and MFP, it became apparent that people are unlikely to learn to operate microcomputers if there are no incentives beyond those inherent in the use itself. This is particularly evident when more is required than simply using menu–driven programs. The ministries' formal systems offered no positive incentives for learning to use the microcomputers or analyzing the information they provide, nor was there any sanction for failing to do so. The incentive systems thus obstructed the use of microcomputers to improve decision–making among the professional staff.

In MOF, Indonesia, there was recognition of the importance of incentives, and the intention to provide financial incentives to those who developed microcomputer skills. In one directorate, selected staff were receiving highly valued training, including training abroad, in various aspects of microcomputer use. These opportunities provided strong incentives for staff to learn to use the microcomputers.

In contrast, informal incentives in MOF discouraged acceptance of the microcomputers by the heads of some local tax and customs offices. Resistance from those offices was considerable. Although effective use of microcomputers in the districts might eventually become established due to the power of headquarters, the process would be greatly facilitated by incentives encouraging the leadership of local offices to accept the microcomputers as a permanent part of the system.

The cases also demonstrate that effective incentives can come from outside the formal system. In MFP and MALD, the perception that microcomputer skills expanded employment opportunities outside government encouraged some ministry staff to learn to operate microcomputers.[13] Advisors introducing the microcomputer system also provide "external" incentives, such as opportunities to study abroad. Effective incentives need not be material, however. Providing an appreciative audience and thoughtful feedback can also served as an influential incentive.

The experiences in Kenya and Indonesia also make it clear that microcomputers are not an apolitical tool. There are likely to be some who perceive their interests to be served by microcomputer adoption, and others whose interests are threatened by adoption. In order for microcomputer use to be established effectively, those who support such use must have control over critical resources.

In the finance divisions of MALD and MFP, Kenya, for example, there was pressure from some quarters to have decisions made on bases other than information. In the absence of countervailing power and incentives, finance officers continued to make decisions on those bases, despite the more accurate and readily available information provided by the microcomputers. In MFP, the budget officers who perceived their authority to be threatened by the microcomputer system strongly resisted the system. In MOF, Indonesia, many district officers considered microcomputers a threat to their power and adamently resisted microcomputer introduction. A critical determinant of the impact of microcomputers in these cases was the relative power of these groups vis–a–vis those supporting microcomputer use, as well as the extent to which the incentive system encouraged ongoing microcomputer use.

All three ministries illustrate the contributions microcomputers can make in improving the availability of accurate information. If the information is to improve the quality of decision-making, however, leadership messages, training in information use, the incentive system, and the power structure must support the change. Yet it is also important to note that there are limits to the extent to which government agencies in any country are willing and able to make decisions solely on a "rational" basis. There is inevitably a conflict between allocating resources efficiently, and responding to other considerations, including political pressure and personal gain.[14] Although microcomputers can attentuate the influence of these "irrational" pressures, they will not eliminate them. Nonetheless, with the proper support from management systems, microcomputers can help significantly to improve the quality of decision-making.

Acknowledgments

This chapter is based primarily on field work conducted August 12–September 12, 1984 in Kenya, and September 13–October 10, 1984, in Indonesia, supplemented by subsequent telephone and personal interviews with advisors working in the ministries. It is important to note that conditions surrounding microcomputer use are dynamic, and that some of the problems discussed here have been identified and are being resolved in the three ministries. I am grateful for the generous assistance of all of the many individuals in Kenya, Indonesia and Cambridge who shared with me their knowledge. Of course final responsibility for the substantive analysis rests with me alone, along with all related shortcomings.

Notes

1. Case studies in LDCs include U.N. Centre for Science and Technology for Development (1985); "The USAID/Honduras Contraceptive Social Marketing Project" (undated: 41–42); Ourusoff (1983); Ingle and Connerley (1984); Bertoli and Bertoli (1981); Mann (mimeo); Pinckney, Cohen, and Leonard (1982); Pinckney, Cohen, and Leonard (1984).

2. See, for example, Allen (1982); Garson (1983); McCaleb (1982); Otway and Peltu (1983); Pava (1983).

3. There is a sizeable literature on innovation in organizations. Regarding the importance of leadership to technological innovation in the private sector, see Cleland (1975). Some of these factors emerged

in computer adoption in U.S. local governments; see Kraemer and King (1977); Danziger, Dutton, Kling, and Kraemer (1982). For a general discussion of some of the important factors in organizational operations that also seem relevant to microcomputer use, see Bower (1983); Jackall; McFarlan and McKinney (1983); Rondinelli (1976); Johnston and Clark (1982).

4. A large literature, particularly about the private sector, notes the importance of positive, visible leadership in legitimating innovation. Albricht (1973), for example, describes the importance of top management backing of a new information system through its "ups and downs." A good discussion of the influence of the manager is in Leonard (1977). Also, regarding the importance of leadership, see Vertinsky and Barth (1972); Guest (1962); Walton (1975). McCaleb (1982) gives practical advice in exercising leadership to the head of a business (or any organization) introducing microcomputers. Mann (1984) suggests that leadership was extremely important in the adoption of microcomputers in the Ministry of Agriculture, Tunisia.

5. See Brodman (1985a); Kraemer and Danziger (1984).

6. Danziger and Kling (1982) argue that in U.S. local governments, decision makers tend to use information from computer systems to buttress their power rather than as a basis for making decisions. Kraemer and Dutton (1982) conclude that computer use in U.S. cities serves those who already have power. Ruttan and Binswanger (1978) are also useful in showing how new technologies generally can affect the distribution of power.

7. The TAP project is conducted by the Harvard Institute for International Development, Harvard University. Discussion of MALD's experience with microcomputers is in Brodman (1985a); Pinckney, Cohen, and Leonard (1984). In 1981, when microcomputers were first adopted, the Ministry of Agriculture and Ministry of Livestock Development were separate. They were united into one ministry, MALD, in 1983.

8. Another important factor motivating the finance officer to learn to operate the microcomputer was the imminent departure of the microcomputer advisor. The finance officers using the microcomputer–produced reports feared that the clerk who could program in dBase II would not be able to take over for the Microcomputer Advisor.

9. RPP is conducted by the Harvard Institute for International Development, Harvard University. Background on the RPP project is discussed in Ministry of Economic Planning and Development (1983) and Saywell (1983).

10. These observations are based principally on field work in August 1984. In telephone conversations in mid-1985, advisors voiced continuing concern about planning officers' limited use of microcomputers for analysis. It is important to recognize that this is a difficult problem in many other contexts, including the United States.

11. The importance of the incentive structure is also demonstrated by the contrast in the responses of finance and planning officers and those of clerical/secretarial staff. Whereas only a few finance and planning officers have learned to use microcomputers—primarily for word processing or simple menudriven programs—more than a dozen secretaries have learned word processing, and a clerk has become a highly skilled programmer in data management software. The clerk and secretaries are responding to incentives external to the formal ministry system. They have received direct financial incentives from the RPP project for learning to operate the microcomputers. They also expect to improve their employment opportunities outside the Ministry by developing microcomputer skills.

12. For detailed discussion of the use of microcomputers in MFP budget production, see Wescott (1986) and Brodman (1985b).

13. It is also important to note, however, that the incentive system should not only encourage people to learn to use the microcomputers, it should do so in ways that support ongoing and stable use. In MALD and MFP, Kenya, for example, external incentives to develop microcomputer skills include job opportunities outside the Ministry. Clearly, such incentives do not support institutionalization of microcomputer skills in the ministry.

14. The problem of getting decision makers to use information effectively as the basis for decisions also occurs in other contexts. See, for example, Evan (1976); Caiden and Wildavsky (1974); Guyot (1977); Danziger and Kling (1982); Pauley and Cohen (1984). On the other hand, Ingle ((1983): 23) found that "once the technology is understood there is a tendency to use it as a solution to every problem encountered, leading in [sic] inefficient use of the machine."

References

Abrecht, L. K. (1973). *Organization and Management of Information Processing Systems*. New York: MacMillan Co.

Allen, Brandt. (1982, November-December). An Unmanaged Computer System Can Stop You Dead. *Harvard Business Review*. 77-87.

Bertoli, F., and Bertoli, S. (1981). *The Utility and Potential Applications of Microcomputers in Data Processing and Analysis: A Re-*

port On Operational Support in Rabat, Morocco (AID/DSPE-C-0053). Washington, D.C.: U.S. Agency for International Development.

Bethke, K.; Cohen, J. M; and Hook, R. M. (1984, June). *History of Rural Development Planning in Kenya*. Mimeo.

Bower, J. L. (1983, July–August). Managing for Efficiency, Managing for Equity. *Harvard Business Review*, 83–90.

Brodman, J. Z. (1985). *Microcomputer Adoption in Developing Countries: Old Management Styles and New Information Systems*. A report prepared for the International Development Management Center, University of Maryland, College Park, Maryland.

_____. (1986a). *Microcomputer Adoption in the Ministry of Finance and Planning, Kenya*. Development Discussion Paper No. 219. Cambridge, MA: Harvard Institute for International Development, Harvard University.

_____. (1986b). *Microcomputers to Improve Decision-making in Third World Governments*. Development Discussion Paper No. 231. Cambridge, MA: Harvard Institute for International Development, Harvard University.

Caiden, Naomi, and Wildavsky, Aaron. (1974). *Planning and Budgeting in Poor Countries*. New York: John Wiley & Sons, Inc.

Cleland, D. I., and King, W. R. (1975). *Systems Analysis and Project Management*. New York: McGraw-Hill Co.

Danziger, J. N., and Kling, R. (1982). Computers in the Policy Process. In *Computers and Politics: High Technology in American Local Governments*. Danziger, J. N.; Dutton, W. H.; Kling, R.; and Kraemer, K. L. New York: Columbia University Press.

Evan, W. M. (1976). *Organization Theory: Structures, Systems and Environments*. New York: John Wiley & Sons.

Garson, G. D. (1983, September–October). Microcomputer Applications in Public Administration. *Public Administration Review*, *43*, (5), 453–458.

Guest, Robert H. (1962). *Organizational Change: The Effect of Successful Leadership*. Homewood, Ill.: The Dorsey Press, Inc. & Richard D. Irwin, Inc.

Guyot, J. F. (1977, November–December). How Do We Know Which Training Is Good Medicine for Managers? *Public Administration Review*, 698–705.

Ingle, M. D. (1983, April). *Evaluating the Appropriateness of Microcomputers for Management Applications in Developing Countries*.

Paper presented at the American Society for Public Administration, National Conference, New York, N.Y.

Ingle, M. D., and Connerley, E. (1984, January). *Microcomputers and Program Management in Portugal: A Case Study of Emerging Technologies in the Agriculture Sector*. Paper prepared for the International Labour Organization, Division of Science and Technology, Geneva.

Jackall, R. Moral Mazes: Bureaucracy and Managerial Work. *Harvard Business Review*, 118–130.

Johnston, B. F. and Clark, W. C. (1982). *Redesigning Rural Development*. Baltimore, Md.: Johns Hopkins University Press.

Kraemer, K. L., and Daniziger, J. N. (1984, January–February). Computers and Control in the Work Environment. *Public Administration Review*, 32–40.

Kraemer, K. L., and King, J. L. (1977, February). *A Critical Assessment of Urban Technology Transfer: The Case of Computing Applications in U.S. Local Governments*. Paper prepared for *Transfer 4*. Irvine, Calif.: The URBIS Group, University of California, Irvine.

Leonard, D. K. (1977). *Reaching the Peasant Farmer*. Chicago: University of Chicago Press.

Mann, Charles K. (1984). *Microcomputers in Agricultural Planning in Tunisia: Report of a Follow-up Visit*. Mimeo.

_____. Beyond the Metaphor: The Microcomputer in Public Policy and Human Capital Development, Chapter 1.

McCaleb, Robert B. (1982). *Small Business Computer Primer*. Beaverton, Ore.: Dilithium Press.

McCalla, A. F., and Saywell, J. T. (1983, October). Mid Term Evaluation Report, Harvard Technical Assistance Pool Project, Phase II. Mimeo.

McFarlan, F. W., and McKinney, J. L. (1983, July–August). the Information Archipelago—Governing the New World. *Harvard Business Review*, 91–99.

Ministry of Economic Planning and Development, Government of Kenya. (1983, January–August). *Rural Planning Project Progress Report*. Nairobi: Government of Kenya.

Otway, H. J., and Peltu, M. (Eds.). (1983). *New Office Technology: Human and Organizational Aspects*. Brussels: Ablex Publishing Co.

Ourusoff, Nicholas. (1983, September). *Planning in Malawi: The National Statistical Office and Ministry of Agriculture*. Mimeo.

Pauley, P. A., and Cohen, S. (1984, April). Facilitating Data-Based Decision-Making. *Evaluating Review*, *8*, (2), 205–224.

Pava, C. (1983). *Managing New Office Technology: An Organizational Strategy*. New York: Free Press.

Perry, J. L., and Karemer, K. L. (1979). *Technological Innovation in American Local Governments/The Case of Computing*. New York: Pergamon Press.

Pinckney, T. C.; Cohen, J. M.; and Leonard, D. K. (1982, August). *Microcomputers and Financial Management in Development Ministries: Experience From Kenya*. Development Discussion Paper No. 137. Cambridge, Mass: Harvard Institute for International Development, Harvard University.

_____. (1984, June). *Kenya's Use of Microcomputers To Improve Budgeting and Financial Management in an Operating Ministry: An Updated Report*. Development Discussion Paper No. 169. Cambridge, Mass: Harvard Institute for International Development, Harvard University.

Rondinelli, D. A. (1976, Fall). International Assistance Policy and Development Project Administration: The Impact of Imperious Rationality. *International Organization*, *30*.

Saywell, J. T. (1983, September–October). *Rural Planning Project, Phase II, Midterm Evaluation for USAID*. Report to USAID, Department of State, U.S. Government.

Schwartz, H.; Bennigson, L.; and Brodman, J. Z. (forthcoming). The CEO's Change Agenda. *Organizational Dynamics*.

The USAID/Honduras Contraceptive Social Marketing Projects, A Critical Path Analysis - Using an Apple II & a Visi-Schedule Program. *Mimeo*.

United Nations Centre for Science and Technology for Development. (1985). Microelectronics and Decentralized Development. *Advanced Technology Alert System Bulletin*, second issue.

Vertinsky, I., and Barth, R. T. (1972). A Model of Diffusion and Implementation: An Exploratory Study of Managerial Innovation, Columbia. *Socio-Economic Planning Sciences*, *6*, 153–171.

Walton, R. E. (1975). The Diffusion of New Work Structures: Explaining Why Success Didn't Take. *Organizational Dynamics*, *3*, (3), 3–22.

Wescott, C. G. (1986, April). *Microcomputers For Improved Budgeting By The Kenya Government*. Development Discussion Paper No. 227. Cambridge, Mass: Harvard Institute for International Development, Harvard University.

8

The Dilemma of Acquiring Micro-Based Software in Developing Countries

Barry K. Render and Stephen R. Ruth

Introduction

Long ago it was predicted that computing power would one day be within the reach of every nation in the world. In its 1970 ten-year plan for developing countries, the U.N. wrote:

> Computers will play an increasingly important role in developing countries which intend to participate in the world economy in ways other than the supply of raw materials. Developing countries will find computers a necessary ticket of admission. The next decade should see developing countries even more active in closing the computer gap.[1]

Unfortunately, the U.N. prediction did not hold true for most developing nations. For every success story in computer technology transfer, such as Taiwan, Hong Kong, Korea, and Singapore, there are ten Pakistans, Nepals, Tunisias, Ethiopias, and Bangladeshs that have encountered severe financial political, cultural, and business opposition and disappointments in attempts to computerize.

Advanced western countries and Japan continue to increase the lead over their less-developed neighbors. They now control a computer capability which, while altering their production and service structure, is also *reinforcing* their traditional advantages in science and technology. Even though developing nations have significantly lower unit labor costs, the proliferation of microprocessors, microcomputers, robotics, and office automation in industrial nations is increasing productivity to the level of countering the wage differential.

What can developing nations do to pull themselves into the twentieth century with respect to computer use? There are, of course, many major problems to be tackled. These include a tremendous shortage of trained computer personnel, poor national planning and prioritizing, minimal national sharing of resources, fear of automation and job loss, and perhaps most importantly, limited foreign reserve currency for hardware and software purchase.

In developing countries, every penny spent on data processing is one less spent for grain imports, consumer goods, tractors, manufacturing equipment, and telecommunications. One possible way for a poor nation to cut its computer bill is local assembly or manufacture of computer hardware. Apart from Japan and the industrialized western nations, India is one of the few countries to attempt to develop a hardware industry. With strict rules governing the import of hardware, India now produces two–thirds of its total internal requirements in mini– and microcomputers. But its industry is a decade behind western technology and is not known for its high quality. Manufacture of DP equipment is a difficult business to enter, requiring billions of dollars of R & D investment which is unavailable to but a dozen or so countries in the world.

This chapter discusses another option—that of tackling software costs. The authors' objectives are to describe the current international software market and to propose a niche and possible strategy for developing nations in that market.

The International Software Market

In 1960, about 90 percent of the data processing budget in a typical organization was devoted to hardware and only 10 percent to software. By 1984, a dramatic change had taken place, with 60 percent or more of that budget now spent on software. Recognizing this, several developing nations have turned to software development as an industry for both internal use and export.

Writing computer software is a *relatively* easy export area open to less-developed countries for two reasons: (1) it has a low capitalization rate (especially compared to hardware production), and (2) it consumes little more than brainpower. Since many nations, such as Pakistan and Egypt, have a continuing brain drain of talented, underutilized college graduates to the United States, other western countries, and Saudi Arabia, a new brain–intensive industry such as software development may make sense to provide more opportunity at home.

India, for example, exported $20 million of software products in 1982, selling to Diner's Club, World Bank, and Ford Motors, among others.[2] In 1983, Nepal opened its first software house, funded by an American industrialist who recognized the plethora of educated analytic minds—and the advantage of a programmer salary structure of $1,000 per year (compared to over $25,000 in the United States and $10,000 in Japan.[3] Also in 1983, the very computer-progressive government of Singapore formulated financial incentives to induce overseas organizations to set up software creation and production facilities on their island. Their plan encouraged foreign entrepeneurs to open joint software houses with local firms: U.S., British, and Japanese vendors have already done so.[4]

The international software market may not be as easy to crack as it seems, however. The basic question is whether Third World countries can develop the expertise to pursue the competition seriously. The competition is tough—and it is mostly American.

The American Software Domination

It took just two years for the International Business Machines Corporation to dominate the personal computer market worldwide as surely as it dominates the micro and mainframe markets in the United States. And while IBM commands the hardware side of the picture, American software companies totally dominate overseas software sales, especially with respect to personal computers. Lotus Development Corp., Ashton-Tate, Microsoft, MSA, and others have taken American versions of their products, tailored them using translators and local distribution networks, and put them on overseas markets before local software firms could economically justify investment. Ashton-Tate for example, derives about 25 percent of its $80 million total sales from the overseas market: it ships 11 different language versions of its products, such as dBase II, abroad. Even in Europe, with a strong tradition of software creativity, especially in England and France, there is concern about the overpowering presence of American software. According to International Data Corp. statistics for 1984, U.S. companies accounted for nearly 87 percent of the personal computer software in France. In Germany, the best a local software company could do was place seventh behind six American firms. And even England pays most of its micro software money to American companies.[5] One way the Europeans propose to catch up in building a software base is by acquiring American software companies that are faltering in the competitive shakeout taking place in the United States.

In addition to a competitive spirit from the United States and Europe, developing countries face a dedicated newcomer in the software market, Japan. In an article entitled "Japan's Push to Write 'World–Class' Software," *Business Week* states:

> Faced with the twin problem of rapidly falling hardware prices and the growing percentage of computer budgets being spent on software, Japanese hardware manufacturers are scrambling to shift their resources into software development.[6]

Five years ago, for example, Hitachi spent 10 percent of its R & D budget on software. Last year, that figure was up to 30 percent. Toshiba recently opened two factories, employing 5000 software engineers, to develop industrial software. NEC Corp. has hired U.S. software engineers to help it crack the huge American market and assist in developing commonly used business applications.

The Japanese learned early that a shortage of applications software severely limited the export potential of Japanese personal computers. They are keenly aware that to continue to be competitive internationally in hardware, they must start developing world–class software. This means selling more software overseas to recover development costs.

The Key to Software Success

Why does the United States have a seemingly impenetrable worldwide lead in the software industry? The answer lies in but two words: *Business Skills*. It is not that there are not talented COBOL or FORTRAN programmers in Sri Lanka, Nepal, or Nigeria. The issue, as Japan is discovering, is that success in software development depends upon a knowledge of business applications in the marketplace. Because of language, cultural differences, and differences in accounting rules, much Japanese software is useless overseas—even *after* it has been translated from Japanese Kanji characters to English or French or Arabic.

Werner Frank, writing in *Computerworld*, recognized this when analyzing Singapore's drive into the software industry. Note his final paragraph of the three quoted below:

> Financial incentives are being formulated to induce overseas organizations to set up software creation and production facilities in Singapore beginning in 1983. However, it should be clear to those interested that the motive is to create a software export business rather than to supply markets in Singapore alone.
>
> If the past progress and determination of Singapore's government are any indication at all, we can expect that the objec-

> tive of making the area a software center will indeed be realized.
>
> But Singaporeans should be aware of one thing: Software innovation is not something that results from government decree. Rather, it springs spontaneously from the brain of human beings who are rich in hands-on application-oriented experience.[7]

Herein lies the danger of such new software firms as the one recently opened in Nepal. The Nepalese are not necessarily developing software export skills of the type Frank refers to above. They are being utilized as a pool of inexpensive programmers to code from systems analysis specifications already developed in the United States or Europe. This industry pays well, but is neither intellectually stimulating or satisfying. (It is almost analogous to the keyboarding for databases of back issues of the *Washington Post* and *New York Times* done in Taiwan and the Phillipines. The word processing clerks do not speak or read English, and are hence less likely to be distracted by the contents of the text they are replicating. They are used because they are hard working, accurate, and cheap.)

The success of an application system depends not on the data processing skills of the programmer/analyst, but on that person's applications knowledge and understanding of the intricacies of each business. The solution for developing nations that wish to conserve on the import of expensive foreign software and to enter the software development business can take on several forms, some short-term and others longer.

Agenda for Action

In the short run, countries can begin to develop expertise in software production and experience in marketing by establishing joint ventures with well-funded western nations seeking *entre* into developing markets. It is still not too late to join this new international industry. Secondly, they can seek to import software *technology*—not just software products. This can be done through careful use of experts from developed countries, and through import of software packages with the intention of learning the technology embodied in the packages. Third, developing countries can *share* more widely software libraries and experiences internally, and among themselves. India and Malaysia are good examples of countries with central planning commissions to monitor and advise on computer use, planning, and procurement.[8,9]

An emerging niche for software development may be found in the diversity of hardware which has been acquired by some of the smaller developing nations. A U.S. Department of Commerce report notes the use of at least 20 different brands of microcomputers in small countries like Mali, Cook Islands, Sierra Leone, and specifically throughout Francophone Africa.[10] The very existence of these less-popular systems is an invitation, it would seem, to develop software for them in a relatively unchallenged environment where large Western software firms may not be interested in the return-on-investment. For example, the country that uses the Wicat System 150 or the Smoke Signal Chieftain 9822 (both computers that are less than household words) may be glad to have a local software team developing applications programs.

In the longer term, the answers lie in more education, for both managers and DP professionals, and the development of industry computer specialists. To produce applications software, be it for banking, sugar refineries, or train scheduling, generalists are not as good as specialists in reducing the time and cost of system development. Finally, as the general knowledge level about data processing in a society as a whole increases, so will the environment for solution of the long-term problem of dependency on the West.

References

1. *The Application of Computer Technology for Development*, United Nations, Department of Economic and Social Affairs, New York, 1971.

2. Render, Barry K. "Introduction of Microcomputers in Nepal," *Computerworld*, June 20, 1983, pp. 50–51.

3. *Ibid.*

4. Frank, Werner L. "Singapore Vying for Software Center Stage," *Computerworld*, March 14, 1983, pp. 47–8.

5. Schrage, M. "IBM Wins Dominance in European Computer Market," *Washington Post*, February 17, 1985, p. E–17.

6. "Japan's Push to Write 'World-Class' Software," *Business Week*, February 27, 1984, pp. 96–98.

7. Frank, *op. cit.*

8. Render, Barry K. "Computerization in South Asia," *Data Processing*, Vol. 25, No. 8, October, 1983, pp. 14–17.

9. "A Look at Computing in Southeast Asia," *Computerworld*, January 31, 1983, pp. 29–30.

10. Diskin, B. *Considerations for Use of Microcomputers in Developing Country Statistical Offices*, U.S. Department of Commerce, Washington, D.C., 1983, chapter 11.

9

Technology Sharing Organizations: A Goal-Oriented Approach to More Efficient Solution of Smaller Scale Problems of Development

Stephen R. Ruth

Introduction

The Difficulty of Delivering a Small Amount of Technology

The basis of technology transfer between nations is the relatively abundant resources for solving technological problems in certain parts of the world and the strikingly large number of such problems needing to be solved in others. Perhaps nowhere can this be seen more vividly than in various large developmental projects where a donor nation will transfer large amounts of technology to a client nation. While these massive transfers are usually thought to be of value at several levels of the client nation as well as the donor nation, frequently a secondary effect is a constellation of unsolved, small, indigenous problems. Yet the technology is needed. Whether viewed from a perspective of total factor productivity[1] or other measurable dimensions, it is clear that underdevelopment can mean vulnerability. Particularly for import-dependent nations, rapid industrialization can be viewed as a cure for many ills, including unemployment problems.[2]

Not surprisingly, then, the smaller, less cosmic types of problems which affect countries are not as likely to be as wholeheartedly supported as the larger issues. Yet in a study by the National Science Foundation[3] which assessed national and international problem areas, the majority of the 41 problems could most typically be classified as amenable to more focused approaches--that is, they were problems which did not require the application of massive resources, very extensive elapsed time for delivery, or disruption of existing organic structures. The position of this chapter is that in attacking problems of smaller size, at a lower level, and using some existing instrumentalities of the world's scientific organizations, it may be possible to obtain solutions which are more efficient, more durable, and possibly better supported by the client region.

Reasons for Difficulties in Dealing with Smaller Development Problems

Client-Centered Difficulties

The reason for the difficulties in being able to deliver small developmental programs are many, but they can be classified briefly in terms of being client-centered, donor-centered, or technology-centered. One of problems which relates to the client or user is the perception that some of the rules of development programs are both frustrating and humiliating. For example, as Myint[4] has noted, there are some types of problem solutions where the span is of too short a term for the World Bank yet too long for the International Monetary Fund to be involved. In the same vein, a review of the Middle East region for nine countries from 1970-1975[5] indicates that a significant amount of the technological transfer from all donor nations was hardware, software, massive training, buying computer networks, airplanes, and the like. More modest, small-scale, regionally centered help, like locust control in parts of Egypt, was not the central or salient theme of most of this developmental aid and this tendency has not changed significantly. The implication is not that such large technology transfer is not useful. But in the process of examining ways to approach aid, donor and client nations may assume, perhaps incorrectly, that bigger is better. Another complicating influence is the tendency of many countries to have relatively fragile planning structures. The examples cited by Fry[6] are indicative, certainly not exhaustive, but do point to further encumberances in the ability to deliver needed small-scale technological help to developing nations.

The perception which the client nation may have concerning the cost of technological intervention involved can frequently be a serious, possibly devastating factor in the total process. Some of the less developed countries in Eastern Europe have shown considerable concern for full pricing of technological aid, reasoning that if the technology is already developed, why should a donor nation charge for it again?[7] Finally, the social context of technology is significant. Values essential to understanding the social context of technology are rather different in socialist systems, for example, than elsewhere.[8] Further, the statistics of technological change, like persons per patent and other such mechanistic indexes, are relatively unavailing when the change involved is perceived as being a donation with too many strings attached. In spite of this, it is beginning to be recognized that the developing countries can gain by profiting from technological errors of donor nations. Waiting for the technology to stabilize, then, is also good national policy.[9]

Donor–Centered Difficulties

Difficulties in delivering small doses of shared help are equally attributable to the misperceptions and inefficiencies of the nations donating the help. Some nations may take the view that "all aid is conceived in support of the donors' interests rather than of the development of the recipients."[10] While this judgment is probably too harsh, it does point out the possibility that aid may be viewed in a context of magnanimity by the donor nation. Further, the intervention or type of improvement offered may be inappropriate for the region. Haberler has commented on the futility of new agricultural technologies when they are not considered in a location-specific context. His comment, "technology often fails when shifted,"[11] is a reminder of the problems donors often force upon themselves in trying to solve such problems.

The donor nation's own perspective of the role of technology can often been misunderstood. Particularly in the developed nations, the dependence on any technology is almost absolute. As Mishan observed, the concept of "growthmania"[12] often leads technologists to solutions which, though appropriate for large systems are sometimes inappropriate for small ones.* In addition, for the developed countries, the intervention of government in technology is often taken for granted to the point where government contacts often lead to radical changes in management practices and technology.[13] This organizational milieu can distort and, in some cases, eventually subvert the traditional donor/client relationship, especially where the client is not accustomed to the government's influence in funding research.

*Incidentally, Mishan suggests that the New York blackout in 1965 may have been a felicitous technological failure since it reminded the citizenery of the fallability of even the most dependable systems.

Technology-Centered Difficulties

Finally, the nature of the technology itself can intervene to complicate a process which is already stretched to a serious limit. Even if problems of client and donor were not involved, that is, if the technology were imported directly by the client with no donor commitments, the potential problems are quite serious. The impact of introducing special engines and diving equipment on the sponge industry in Kalymnos, Greece is a microcosm of the problems which occur when a small, regionally centered technology is challenged anywhere.[14] Schumacher[15] has eloquently described the importance of delivering technology which is useful and efficient and which does not disrupt dramatically the ambiance of the locale.

The problems are significant indeed. Whether one's perspective is the client, the donor, or the technology itself, the difficulties are daunting. Yet there may be a procedure for delivering small, efficient doses of sharing help to nations and regions in ways which defy the odds and by any estimate the odds against their successful implementation of informatics are high indeed.[16,17]

Technology Sharing Organizations: Harnessing Existing Scientific Organizations for Smaller Problems

In an effort to capitalize on existing structures in donor as well as client nations, it may be possible to devise efficient delivery techniques which stress regional, functional sharing and which deliver solutions relatively rapidly. The Technology Sharing Organization (TSO) approach is dedicated to the concept that for modest sized problems it can be possible to develop solution options in a relatively short time span and simultaneously to assure relatively rapid completion of the solutions which are developed.

Table 9.1

Problem Types Which Could Utilize the TSO Approach

Data communications-networks for businesses, law offices, etc.
Certain generic small business problems
Language problems (fluency)
Automation literacy
Computer literacy
Library sharing arrangements
Shared data banks (insurance, law, etc.)
Agricultural knowledge sharing (especially regional pesticides and fertilizers)

Table 9.1 gives an example of some of the important problem types whose solutions might yield high return in a specific region. As an example, automation literacy can be a large national policy issue, or it can be examined in a much more focused context. Many cities, counties, and towns have found that the advantages of automation are seldom achieved without a general Lewinian "unfreezing" of potential users prior to delivery of such systems. Therefore, automation literacy, like library–sharing arrangements and insurance data banks, might be a paradigm of small, focused problems which, when articulated by specific regions, could be capable of a more direct and efficient approach for solution.

TSO Problem Solving Approaches.

The TSO could utilize many possible intervention techniques. Table 9.2 gives examples of some of these modalities. In each case, the advisors are persons who are expert in their field, comfortable with the region involved, and ready to develop problem definitions and potential solution descriptions in a short time.

Table 9.2
Examples of Possible TSO Intervention Methods

Recommending proven training techniques
Establishing specific control modalities
Focusing broad problems toward existing technologies
Management advising or consulting
Referral to other qualified TSO advisors
Facilitating local or regional sharing of specific computer hardware or software

While there are obvious advantages to deep and lengthy perusal of a problem, quite frequently the use of experts who are carefully prepared before their visit and well focused during their visit can permit the maximum use of the techniques like those shown in Table 9.2.

The TSO Sequence

For the TSO approach to be effective, a fundamental requirement would be that an agreed–upon process and procedure be organizationally mandated. Table 9.3 gives an example of such a process. It describes in a general way the sequence of eight months between problem identification and actual generation of solutions and feedback mechanisms for final implementation. It should be noted that the procedure involves careful, early planning for visits and goal–oriented discussions with all members of the group. Further, it forces on the client as well as the TSO representatives a responsibility for developing very rapid and clearly thought-out options. The nature of the problems which can use this approach should dictate a process which is both technically feasible and administratively workable.

Table 9.3
General Sequence for TSO Approach

Time	Action	Responsibility	Comment
X	Identify possible problem areas and communicate with TSO	Local or Regional Client	Problem statement is specific; should identify the sector, functions and key issues.
X+1 month	TSO Representative(s) briefly visit(s) site for initial problem definition.	TSO	Visit is a week or less. Result is brief, agreed-upon description of problems and possible intervention techniques.
X+2 mos.	TSO develops more detailed problem description; plans individual or team visit.	TSO	This step may not be needed if previous step is able to give needed results.
X+3 mos.	TSO visit and development of detailed problem statement and proposed solutions.	TSO/ Client	A formal statement is prepared.
X+4 mos.	Agreement reached on statement and solution. Specific assignment of further TSO help made.	TSO/ Client	Agreements by all parties of what must be done and initiation of work.
X+8 mos.	Solution completed. Feedback mechanisms for continued resolution left in place.	TSO	Most tasks will be relatively short duration.

The TSO Approach: An Example

Table 9.4 gives a hypothetical example of an actual case where the TSO would be brought to bear on existing problems. In this case, the client region would be the city of Alexandria, Egypt, and the intervention technique would be several visits of skilled TSO representatives to Alexandria. The entire process requires eight months from description of the initial perception of need to solution options and completed feedback mechanisms and schedules.

Administering TSO - A Simple Efficient Approach

The TSO, since it is primarily a clearing house, easily could be administered with a very small staff at a university or perhaps through some non-profit organization. The initial task of selecting appropriate TSO organizations and individuals could be coordinated through agencies', societies', and other organizations' periodic meetings as an appropriate agenda matter. Once established, the TSO would be able to connect the regional problems with the team of capable, highly qualified consultants. TSO would be funded primarily through the clients' payments for services and possibly by occasional grants from member organizations.

TSO: An Approach, Not a Reality—Yet

This proposal for technological sharing organizations is meant to be a prospect, not a specific, inflexible approach. It has been noted that the problems which attach to very large donations of technology, to the developing nations especially, are characterized by lengthy planning cycles, frequent changes and, often, by frustrating, agonizing delays. This approach is designed to respond to a different kind of problem for a smaller region. It seems to have the advantage that organizations already in place in many nations would have the ability to supply the capable help and the initiative to develop problem descriptions and solution sets which are rational, usable, and efficient.

For the client nation, too, the advantages would seem to be congruent. First of all, a smaller region and a smaller problem are involved. Secondly, the solutions and structures are agreed upon throughout the process. Thirdly, the feedback process left in place is not designed for some outside nation or agency's use. It is specifically tailored and agreed upon by the client region and therefore has the potential for being fully implemented and enforced by them.

This proposal does not solve large problems. Yet it may serve as a beginning for other attempts to solve smaller problems at lower levels. As a modest proposal to unite the capabilities of existing scientific organizations with the need for some regions to solve smaller-scale development problems, it would seem that the TSO would be worth a try.

Table 9.4
Hypothetical Example of the TSO Approach to a Realistic Problem

Action		Comments
6/87	Egyptian organization identifies a concern to have more trained system designers for developing business-related procedures in white collar areas like finance and inventory management. Letter to TSO defines the problem and focuses on the region (e.g., Alexandria, Egypt).	TSO has already selected potential specialists, available during specific time periods, awaiting the need.
7/87	Two TSO representatives are selected and spend a week in Alexandria. Before returning, they describe in writing their assessment of the problem in its *in situ* context, and briefly outline what their general solution options will be. Letter from TSO forwards these ideas to client within 2 weeks after visit.	TSO representatives are highly qualified and capable of developing problem descriptions. They have prepared themselves beforehand, so their visit is action-oriented and efficient.
8/87	Because of problem's scope and need for further review, TSO representatives return for another week. Result, before departing, is a detailed problem statement and a set of agreed-upon solution options which include statements of technology, time, and cost.	This return visit will not always be required before final problem statement, but is necessary in this case due to the variety of possible solution options.
9/87	Agreement reached on all specifics of problem definition and specific solution options proposed. Also standards, deadlines, and responsibilities are formalized. In the case of this specific problem, a likely solution set would include short courses in system analysis using programmed instruction, computer-assisted instruction, mini-courses, establishment of training agreements with local colleges, MBA programs, etc.	This statement be comes a "contract" to complete phases in a relatively short time (4 months). Since TSO personnel who have been sent are experts in the problems of systems analysis, the contract will include their best state-of-the-art recommendations.
1/88	Solutions options are completed. Feed-back mechanism and schedules are left in place. In this case, perhaps educational strategies have been found and a number of new goals are left with client to develop.	A PERT or MBO network is established for client's use in following up on needs.

Notes

1. Mabro, R., and Radwin, S. *The Industrialization of Egypt 1939–1973*. (Oxford, Clarendon Press, 1976).

2. Bessant, J. "Information Technology and the North South Divide" in Burns, A (ed.) *New Information Technology,* Ellis Howard, 1984.

3. National Science Foundation, *Assessment of Future National and International Problem–Areas*. Volume 1. SRI Project 4676, February 1977, pp. 83–92.

4. Myint, H. *The Economics of the Developing Countries*. (New York, Praeger, 1965).

5. *Technology Transfer to the Middle East O.P.E.C. Nations and Egypt 1970– 1975*. Committee on Science and Technology, U.S. House of Representatives, September 1976.

6. Fry, James. *Employment and Income Distribution in the African Economy*. London. Croom, 1979.

7. Hayden, E.W. *Technology Transfer to East Europe*. (New York. Praeger, 1976.)

8. Dunn, W.N. *The Social Context of Technology Assessment in Eastern Europe*. In F.J. Fleron, (ed.), *Technology and Communist Culture*. (New York. Praeger, 1977.)

9. Sharif, N. "A Model of Technological Cooperation for Effective Transfer of Computer Technology," in Bennett, J. and Kalman, R. (eds.) *Computer in Developing Nations* (Proc. IFIP Seminar, Melbourne, Australia, October 1980, North Holland 1981).

10. Arnold, G. *AID in Africa*. London. Nichols Publishing Co., 1979.

11. Haberler, G. *Critical Observations on some Current Notions in the Theory of Economic Development*. W. Johnson, (ed.), *Readings in Economic Development*. Southwestern 1972.

12. Mishan, H.J. *Technology and Growth*. (New York. Praeger, 1970.)

13. Black, G. "The Effect of Government Funding in Commercial R&D." Gruber, W.H. and Marquis, G. *Factors in the Transfer Of Technoloqy*. Cambridge, Mass. MIT Press, 1969.

14. Bernard, H., and Pelto, P., *Technology and Social Change*. New York, MacMillan 1970.

15. Schumacher, *Small Is Beautiful: Economics as if People Mattered*. Putnam, 1973

16. Slamecka, V., "Information Technology in the Third World," *Journal of the American Society for Information Science,* Vol. 36, No. 3, 1985.

17. Neelemagen, A. and Tocallian, J. "International Cooperation in Information Systems and Services," *Journal of the American Society for Information Science,* Vol. 36, No. 3, 1985.

Microcomputers in Development: Concluding Observations

Charles K. Mann

The essays in this volume record impressive applications of microcomputers to the problems of the developing countries. As nations accumulate experience, their use of the microcomputer will become increasingly ambitious. Tasks running on mainframes in industrial countries will be attempted on microcomputers in the developing countries. On the one hand, these nations lack the installed mainframe base; on the other, today's microcomputers rival yesterday's mainframes in computer power. Moreover, higher–level languages such as dBase II ease the programming skills bottleneck, making the technology relatively accessible.

Low hardware cost means that redundancy in equipment is a feasible strategy to minimize downtime. Many repairs are simply a matter of plugging in new modules. Power service problems remain, but solutions are becoming more well known.

A dramatic example of how the enormous power of microcomputers is being harnessed by developing country governments is Kenya's use of IBM PCs and dBase II to assemble and produce its entire national budget. Westcott's description of the process suggests that the microcomputer represents an appropriate technology, well–suited to developing country conditions, not something useful only in the industrial West. The other applications reported in this volume provide further support for this technology's suitability.

An exciting aspect of these developments is the quantum leap this equipment promises in giving decision makers more timely and more complete information upon which to make decisions. As noted in the introduction, what is largely a quantitative change in the industrial nations is a qualitative change in the developing countries. Powerful computers are replacing time–consuming, error–prone pencil and paper systems. As experience and confidence grow, users will begin to discover altogether different ways of doing things, as they break through old constraints of manual information processing. Among other things, looking ahead becomes easier. The implications can be seen more clearly of current trends in population, in education, in food demand/supply balances. Early warning systems become feasible for all manner of potential natural or human emergencies. Agriculture recommendations can be tailored more precisely to conditions of a particular locale, even taking into account how the weather and other conditions have been in that year.

Despite the bright promise, not all the signs are positive. While it is too early to judge the effects of widespread microcomputer adoption employment and equity considerations, there is concern both about possible labor displacement and about the development of increasingly bi–modal societies where computer power accentuates the power and priviledge of the dominant groups. Concern has been voiced, too, that women will become more disadvantaged. Were women to remain in their current office worker jobs, this new technology could dramatically raise their productivity. However, once the jobs become technologically enriched, they may cease to be seen as "women's work." Brodman reports that she observed several such instances during her field work in Kenya.

The essays of this volume do not come to grips with these important social issues. Perhaps it is too early in the microcomputer product cycle to do so adequately at this point. As the installations swell, however, research should move beyond documenting particular applications to focus more on the deeper forces at play. These include such questions as: Do computers change the way people think about problems? Is this all to the good and, if not, how can the adverse consequences be mitigated? Do microcomputers tend to centralize or decentralize power? (The evidence of this volume is mixed on that point.) Is the technology labor displacing? Is it gender neutral or adverse to women? What kinds of power shifts will it precipitate, what kinds of organizational structures will it foster? How can the developing countries share more in the generation and production of this information technology? Ruth and Render's chapter suggests some possibilities, but there is scope for yet more work.

One thing that this volume documents is that adoption of microcomputers in the developing countries is moving rapidly, probably far more so than generally appreciated in the West. As prices fall and the advantages of computer power become evident to more people, this trend seems likely to accelerate.

"Microcomputers in Development" already has faded away as a topic which excites donor interest; now they are treated much as any other pieces of office equipment. In some sense, this is probably as it should have been all along. On the other hand, the microcomputer is not just another generation of office equipment. It is fundamentally different and will reshape societies at least as much as the automobile reshaped America. Just at the time administrators have come to accept microcomputers as normal office machines, the first microcomputer-based expert systems are being announced, and huge investments are being made in "artificial intelligence." The technology is outracing the metaphors people apply to it. If the new metaphor is not the human himself/herself, just as surely it is not the typewriter, the magic spreadsheet, or the super filing cabinet.

If there is one overriding weakness in the excellent essays in this volume, it is that they are largely retrospective in documenting particular instances of technological diffusion. They barely scratch the surface of issues which soon will be upon us, not the least of which is the totally different kinds of applications which the next generation of software will permit.

These essays help us to see where we have been. We need now an imaginative and creative look at where we are going and the problems and opportunities we may face on the way. Allowing "Microcomputers in Development" to recede as a topic will not generate the needed vision. It was the metaphors we brought to microcomputers that helped in their rapid diffusion. Ironically, it is these same metaphors that now shackle our thinking about what the next generation of microcomputers is likely to do both for and to the development process.

The real promise of the microcomputer is to be realized not in automating what already is being done, but in making possible things previously unimagined. Enormous contributions to human welfare lie in these realms beyond the metaphor. In the field of development, too few are exploring this uncharted territory.

About the Contributors

Janice Zarro Brodman is Professional Associate at Management Analysis Center, Cambridge, Massachusetts. She wrote her chapter for this book while a Visiting Scholar at the Harvard Institute for International Development where she conducted research on the impact of microcomputer use in developing countries. She has served as a consultant on the topic of the impact of technological and economic change on business and on training and education, particularly in developing countries. She has previously directed various projects studying the relationship of technical applications to employment, education, and public policy. Brodman has published and lectured widely on the subjects of rural development, technological and economic change, and policy making. She received her Ph.D. from Harvard University and a B.A. from Rutgers University and has been awarded various fellowships and honors.

John M. Cohen is Institute Fellow at the Harvard Institute for International Development and Lecturer in the Department of Government at Harvard University. His areas of expertise are rural sociology and development, local agriculture and government, and land tenure and natural resource conservation. Previously, he has served as Assistant Professor in the Department of Rural Sociology at Cornell University, as legal and administrative aid in Ethiopia with the Peace Corps, and in other legal and planning capacities. Cohen has been awarded various fellowships and honors; he has lectured and written extensively on rural development and agriculture; and he has acted as consultant on a number of international technological, economic, and rural planning projects. He earned his Ph.D. in Political Science and B.A. from the University of Colorado, a Juris Doctor degree from University of Michigan Law School, and trained in African legal systems and administration at the University of California, Los Angeles.

Carl H. Gotsch is Associate Professor of Economics at the Food Research Institute at Stanford University. He specializes in agricultural development policy, project analysis, and mathematical programming. Previously, he served as Project Specialist for Agricultural Economics in the Middle East Office of the Ford Foundation, and before that was also a member of the Harvard University Department of Economics. Gotsch has conducted studies and served as consultant on international agricultural issues for various foundations and government agencies. He also has participated in special missions to evaluate and assist agricultural development in India and Egypt. He received a Ph.D. in Economics from Harvard University and an M.S. in Agricultural Economics and a B.S. in Agriculture from Iowa State University. He has published extensively in the areas of international agriculture, economics, and development.

David K. Leonard is Associate Professor of Political Science and Chairman of the Center for African Studies at the University of California, Berkeley. He has lectured and published extensively on the topics of decentralization, government administration, and public policy. He has conducted research and acted as consultant for a number of projects relating to administration, policy making, agriculture, and rural development in Africa. Leonard received a Ph.D. and M.A. in Political Science from the University of Chicago and a B.A. from Haverford College.

Charles Kellogg Mann is Research Associate at the Harvard Institute for International Development. He previously served as Associate Director for Agriculture and Social Sciences at The Rockefeller Foundation. Before that he worked for seven years in Turkey with the RF and the U.S. Agency for International Development; earlier he worked with the U.S. Bureau of the Budget. Mann earned his Ph.D. in Economics and M.P.A. from Harvard University and B.A. from Williams College. His dissertation on technological change received an award "For Professional Excellence" from the American Agricultural Economics Association. He has contributed to and co-edited with Barbara Huddleston *Food Policy: Frameworks for Analysis and Action* and with Malcolm McPherson written a monograph on improving the management of national food systems, with some reference to the role of computers. In addition to articles on economic development, Mann is the editor of a series of audiotapes featuring programs presented at The Boston Computer Society and The Computer Museum.

Philip F. Palmedo is chairman and Chief Executive Officer of the International Resources Group Ltd. He is particularly active in the management of international projects in energy policy and technology transfer. He is involved in assistance with computer applications and economic planning to developing countries and has acted as consultant to a number of national and international institutions and governments. Previously, Palmedo headed the Energy Policy Analysis Division of the Brookhaven National Laboratory. He earned Ph.D. and M.S. degrees from the Massachusetts Institute of Technology and received a B.A. from Williams College. He is author of numerous papers and reports on the topics of energy policy, technology transfer, and computer applications.

Thomas Cotesworth Pinckney, III is an agricultural economist with the International Food Policy Research Institute. His research focuses on agricultural policy, nutrition, and links between advances in information technology and decision making. As Management Advisor for the Harvard Institute of International Development, he has previously been involved in the planning and management of technical applications to food and commodity price and production. Pinckney has published papers on the relationship between international agriculture policies, financial management, and the use of microcomputers. He earned his Ph.D. and M.A. from the Food Research Institute at Stanford University and a B.S. from Davidson College and has received several academic honors and awards.

Barry K. Render is Professor and Chairman of the Department of Decision Sciences at George Mason University. His area of expertise is quantitative business analysis, and he has previously served as Vice President of the Southeast Region of the Decision Sciences Institute. He is a member also of the Institute for Management Sciences, and among the honors he has received are a NASA Federal Fellowship and a senior Fulbright scholarship to Nepal. He has authored five textbooks on management science, operations management, and computers, and he has written numerous articles including several dealing with computer applications in developing countries.

Stephen R. Ruth is Professor of Decision Sciences at George Mason University, where he has also been Lecturer in Management. Previously, he served as Visiting Associate Professor of Business Administration at Georgetown University and as Adjunct Professor of

Information Systems at The George Washington University. From 1955 to 1978, Ruth held senior positions in management at the Department of Defense in various information systems areas. He holds a Ph.D. in Business/Applied Economics from the Wharton School of Finance and Commerce at the University of Pennsylvania; his B.S. was conferred by the U.S. Naval Academy. Ruth consults with a number of corporations and government agencies in the U.S., Europe and South America. In 1985 he was a senior Fulbright lecturer in Argentina, where he set up a major university-based program for training humanities faculty in the use of microcomputers. Among other honors he has received the Defense Department Superior Service Medal, and he continues to serve as Vice President of the American Society of Cybernetics.

Robert Schware is International Affairs Fellow at the Council on Foreign Relations. He is a specialist in the areas of technology transfer to developing countries and management information systems. Papers he has published have concerned the role of microcomputers in developing countries and their use as farm management tools in such countries. Schware received his Ph.D. from the London School of Economics and Political Science, and an M.A. from the University of Essex and B.A. from the University of Colorado.

Alice Trembour is an independent writer in the fields of development and health. In addition to their chapter in this book, she has collaborated with Schware on several other articles about the limitations of microcomputers in developing countries and co-authored with him a popular book on the spreadsheet program Lotus 1-2-3. She received her B.A. from the University of Denver.

Clay G. Wescott, as President of Wescott Associates, is involved in industrial, technical, and investment planning for developing countries. He has been Management Advisor with the Harvard Institute for International Development, supervising the applications of microcomputers and other information technology to financial management systems in Kenya. As Senior Economist for Dames and Moore, Wescott developed international research, tourism marketing, and investment strategies. He earned his Ph.D. from Boston University and received his A.B. in Government from Harvard University. He has published papers on technical, financial, and policy planning in developing countries.

Index